ENGINEERING PHYSIOLOGY

ENGINEERING PHYSIOLOGY

Bases of Human Factors/Ergonomics

Third Edition

K.H.E. Kroemer
H.J. Kroemer
K.E. Kroemer-Elbert

VAN NOSTRAND REINHOLD
I(T)P® A Division of International Thomson Publishing Inc.

New York • Albany • Bonn • Boston • Detroit • London • Madrid • Melbourne
Mexico City • Paris • San Francisco • Singapore • Tokyo • Toronto

Cover design: Greg Simpson

I(T)P®   An International Thomson Publishing Company
         The ITP logo is a registered trademark used herein under license

Printed in the United States of America

For more information, contact:

Van Nostrand Reinhold   Chapman & Hall GmbH
115 Fifth Avenue   Pappelallee 3
New York, NY 10003   69469 Weinheim
                                         Germany

Chapman & Hall
2-6 Boundary Row   International Thomson Publishing Asia
London   221 Henderson Road #05-10
SE1 8HN   Henderson Building
United Kingdom   Singapore 0315

Thomas Nelson Australia   International Thomson Publishing Japan
102 Dodds Street   Hirakawacho Kyowa Building, 3F
South Melbourne, 3205   2-2-1 Hirakawacho
Victoria, Australia   Chiyoda-ku, 102 Tokyo
                                         Japan

Nelson Canada
1120 Birchmount Road   International Thomson Editores
Scarborough, Ontario   Seneca 53
Canada M1K 5G4   Col. Polanco
                                         11560 Mexico D.F. Mexico

1 2 3 4 5 6 7 8 9 10 MVBMP 01 00 99 98 97

Library of Congress Cataloging-in-Publication Data

Kroemer, K. H. E.
    Engineering physiology : bases of human factors/ergonomics /
K.H.E. Kroemer, H.J. Kroemer, K.E. Kroemer-Elbert. — 3rd ed.
        p.   cm.
    Includes bibliographical references and index.
    ISBN 0-442-02380-4 (hardcover)
    1. Human engineering.   2. Anthropometry.   3. Human mechanics.
4. Human physiology.   I. Kroemer, H. J. (Hiltrud J.)   II. Kroemer-Elbert,
K. E. (Katrin E.)   III. Title.
TA166.K76   1997
620.8'2—dc21   97-20681
                                                   CIP

http://www.vnr.com
product discounts • free email newsletters
software demos • online resources

email: info@vnr.com

A service of I(T)P®

Contents

Preface to the Third Edition

The second edition of *Engineering Physiology,* published in 1990, has served its purpose: to provide physiological information that engineers, designers, and managers need in striving to make work and equipment "fit the human." New knowledge of human factors has been gained since 1990, and new ergonomic applications have been developed. Hence, this update. New material in this Third Edition relates especially to:

Chapter 1: Anthropometry and applications of body size information
Chapter 2: Skeletal system
Chapter 3: Muscle and strength exertions
Chapter 5: Biomechanics and modeling of the body
Chapter 8: Metabolism and work
Chapter 11: A new section on "reengineering" deteriorated and damaged body parts.

We appreciate your comments, which tell us what we did well and what we should do better, and especially what to include in the future. You can contact us at P.O. Box 3019, Radford, VA 24143-3019, USA; Telephone (540)639-0514; Fax (540)231-3322; E-Mail karlk@vt.edu.

Karl H. E. Kroemer
Professor, Virginia Tech, Blacksburg VA
Hiltrud J. Kroemer
President, Ergonomics Research Institute Inc., Radford VA
Katrin E. Kroemer-Elbert
Sr. Research Engineer, Howmedica Inc., Rutherford NJ

Preface to the Second Edition

Gunther Lehmann published his *Practical Work Physiology* (in German) in 1952. He used to tell with a smile that his learned colleagues accused him of overly simplifying a difficult subject matter, while many engineers and managers were still baffled by the complexities of the human body. Yet, his book was translated into French, Italian, Spanish, and Polish, and appeared in its second German edition in 1962.

In 1986, we received similar comments on the first edition of our *Engineering Physiology*. The acceptance of the first edition, and its widespread use in courses for engineers and managers have prompted publication of the second edition in 1990.

This second edition is extensively revised. It contains new information on dynamic muscle strength and a chapter on circadian rhythms and work design. But its original purpose is unchanged: to provide physiological information that engineers, designers, and managers need.

We hope to have found a suitable compromise between the necessary depth of information (more can be found in many excellent physiology books) and the desired simplicity and clarity so that the information is "practical."

We thank Ms. Sandy Dalton who expertly and patiently processed this new manuscript.

Radford, VA

Karl H. E. Kroemer
Hiltrud J. Kroemer
Katrin E. Kroemer-Elbert

Chapter 1

Engineering Anthropometry

OVERVIEW

Anthropometric information describes the dimensions of the human body, usually through the use of bony landmarks to which heights, breadths, depths, distances, circumferences, and curvatures are measured. For engineers, the relationship of these dimensions to skeletal "link-joint" systems is important, so that the human body can be positioned relative to workstations and equipment. Most available anthropometric information relies on data taken on military and some civilian populations and measured with classical instruments.

People come in a variety of sizes, and their bodies are not assembled in the same proportions. Thus, fitting equipment to the body requires careful consideration; design for the statistical average will not do. Instead, for each body segment to be fitted, the dimension(s) critical for design must be determined. A minimal or a maximal value, or even a range, may be critical. Often, a series of such decisions must be made to accommodate body segments or the whole body by clothing, work space, and equipment.

THE MODEL

Primary dimensions of the human body are measured between solid identifiable landmarks on bones. Statistical relations among these dimensions, along with others that measure contours, shapes, volumes, and masses, help to describe the body.

INTRODUCTION

The dimensions of the human body and its segment proportions have always been of interest to physicians and anatomists, to rulers and generals, to artists and philosophers, and certainly to anyone who designs or provides any artifact for human use. Marco Polo, writing about his travels around the earth at the end of the 13th century, evoked particular interest with his descriptions of the various body sizes and builds that he saw in different races and tribes. The beginnings of physical anthropology as a recording and comparing science are often traced to his travel reports. Johann F. Blumenbach's 1776 *On the Natural Differences in Mankind* contains all the anthropometric information available up to his day. Alexander von Humboldt encompassed all scientific knowledge in his widely read five-volume *Kosmos,* published from 1845 to 1862.

From about the middle of the 19th century, anthropology became diversified into special branches. Adolphe Quételet, in the mid-1800s, applied statistics to anthropological information. Paul Broca made extensive studies on the skull and drew far-reaching conclusions from them. Biomechanics was an emerging science at the end of the 19th century (see Chap. 5). The rapidly increasing diversity in anthropometric studies led to conventions of physical anthropologists (1906 in Monaco and 1912 in Geneva), who agreed on standards for anthropometric methods. Rudolf Martin's *Lehrbuch der Anthropologie,* first published in 1914, quickly became accepted as the authoritative text and handbook and remained so for many decades. New engineering needs and developing measuring techniques, as well as advanced statistical techniques, gave reasons for updating and redirections in the 1960s and 1970s; under H.T.E. Hertzberg and C.E. Clauser, the Anthropology Branch of the U.S. Air Force was a driving force in anthropometric research and engineering applications. Important compilations of anthropometric data, techniques, and methods were published during this period in the United States (Chapanis, 1975; Garrett and Kennedy, 1971; Hertzberg, 1968; NASA, 1978; Roebuck, Kroemer, and Thomson, 1975).

Measurement Techniques

The classical measuring technique employed four ways to measure stature: with the subject standing naturally upright but not stretched; standing freely but stretched to maximum height; leaning against a wall with the back flattened and stretched to maximum height; and lying in a supine position. The differences between the measures when the subject stretches to maximum height either standing freely or leaning are within 2 cm. Lying face up results in a taller measure, but standing "slumped" reduces stature by several centimeters. This example shows that standardization is necessary to assure uniformity in postures and results.

Terminology and Standardization

Body measurements are usually defined by the two end-points of the distance measured, such as elbow to fingertip; stature starts at the floor on which the subject stands and extends to the highest point on the skull.

The following terms are used in anthropometry.

Height is a straight-line, point-to-point vertical measurement.

Breadth is a straight-line, point-to-point horizontal measurement running across the body or a segment.

Depth is a straight-line, point-to-point horizontal measurement running fore and aft.

Distance is a straight-line, point-to-point measurement between landmarks on the body.

Curvature is a point-to-point measurement following a contour; this measurement is neither closed nor usually circular.

Circumference is a closed measurement that follows a body contour; hence, this measurement usually is not circular.

Reach is a point-to-point measurement following the long axis of an arm or leg.

In general, descriptions of anthropometric measurements include at least three different types of terms. The *locator* identifies the point or landmark on the body whose distance from another point or plane is being measured. The *orientator* identifies the direction of the dimension. The *positioner* designates the body position that the subject assumes for the measurement, such as standing or sitting. Appendix B at the end of this chapter lists many of the terms used in anthropometry.

Traditionally, anthropometric measures, such as those previously mentioned or weight or volume, are taken and reported in the metric system. Professional physical anthropologists are trained to take such measurements; however, with standardization, and with training and supervision by an experienced measurer, other nonspecialists have successfully taken many measurements. As newer measuring techniques evolve (see section "New Measurement Methods" in this chapter), the demands on both measurers and measuring procedures are changing.

For most measurements, the subject's body is placed in a defined upright straight posture, with body segments at either 180, 0, or 90 deg to each other. For example, the subject may be required to "stand erect; heels together; buttocks, shoulder blades, and back of head touching the wall; arms vertical, fingers straight. . . ." This is close to the so-called anatomical position used in anatomy. The head is positioned in the Frankfurt plane, with the pupils on the same horizontal level; the right tragion (approximated by the ear hole) and the lowest point of the right orbit (eye socket) are also placed on the same horizontal plane. When measures are taken on a seated subject, the (flat and horizontal) surfaces of seat and foot support are so arranged that the thighs are horizontal, the lower legs vertical, and the feet flat on their horizontal support. The subject is nude, or nearly so, and unshod.

Figure 1-1 shows reference planes and descriptive terms often used in anthropometry. Figure 1-2 illustrates important anatomical landmarks of the human body in the sagittal view, while Figure 1-3 shows landmarks in the frontal view. Figure 1-4 indicates the postures that are typically assumed by the subjects for anthropometric measurements. Publications by Garrett and Kennedy (1971), Gordon et al. (1989), Hertzberg (1968), Lohman, Roche, and Martorell (1988), NASA (1978),

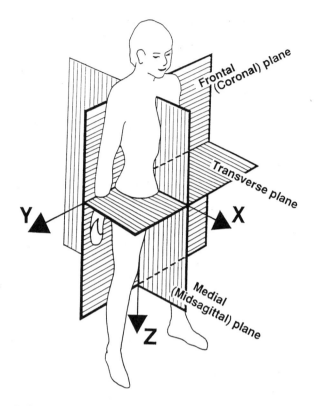

Figure 1-1.   Reference planes used in anthropometry.

Roebuck (1995), and Roebuck, Kroemer, and Thomson (1975) contain illustrations, descriptions, and definitions of landmarks and measuring techniques.

Classical Measuring Techniques

In conventional anthropometry, the measurement devices are quite simple. The Mourant technique uses primarily a set of grids, often attached to a corner of two orthogonal vertical walls. The subject is placed in front of the grid, and the projections of body landmarks to the grid are used to determine their values (see Fig. 1-5). Other setups may include boxlike jigs that provide references for measurements of head and foot dimensions (see Fig. 1-6 and 1-8).

Many body landmarks, however, cannot easily be projected onto grids. Here, special instruments are available, some of which are shown in Figure 1-7. The most important is the anthropometer, which consists basically of a graduated rod with a sliding branch at a right angle to the rod. This sliding branch allows the measurer to reach behind corners and folds of the body, whose distances from the reference are read from the scaled rod. The rod itself can usually be sectioned for ease of transport and storage. The anthropometer has one branch permanently

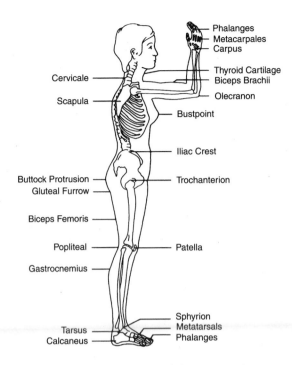

Figure 1-2. Anatomical landmarks in the sagittal view.

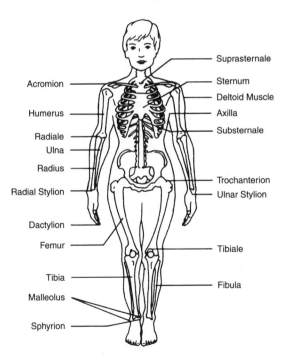

Figure 1-3. Anatomical landmarks in the frontal view.

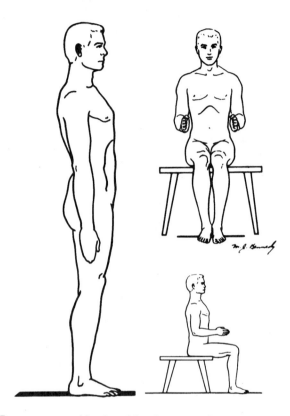

Figure 1-4. Postures assumed by the subject for conventional measurements.

affixed to its top part. This allows distance measurements between the fixed and the sliding branches; this setup is called a sliding caliper, or compass.

The spreading caliper consists of two curved branches joined in a hinge. The distance between the tips of the two opposing ends is read on a scale that is attached in a convenient place between the two parts of the caliper. Many other devices are also employed. A small sliding caliper serves for short measurements, such as finger thickness or finger length; a special caliper measures the thickness of skinfolds; a cone is used to measure the diameter around which fingers can close; circular holes of increasing sizes drilled in a thin plate indicate external finger diameters.

Circumferences and curvatures are measured with tapes, usually made of flat steel, occasionally of nonstretching woven material. Figures 1-8 to 1-10 show typical uses of anthropometer, caliper, and tape, respectively. Scales, of course, are used to measure the weight of the whole body or of body segments. Many other measuring methods have been applied in special cases, such as the shadow technique, templates, multiple probes, and casting. These are further detailed by Gordon et al. (1989), by Roebuck (1995), and by Roebuck, Kroemer, and Thomson (1975).

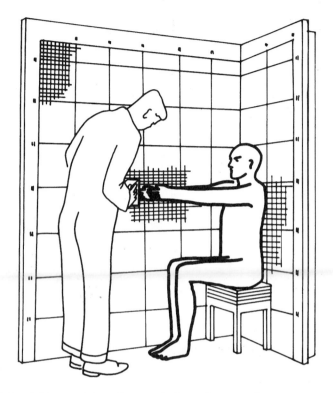

Figure 1-5. Grid system placed in a corner for anthropometric measurements (adapted from Roebuck, Kroemer, and Thomson, 1975).

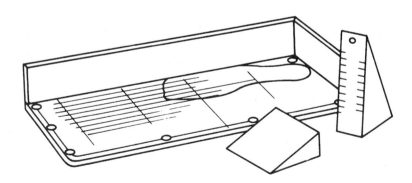

Figure 1-6. "Measuring box" for foot measurements (adapted from Roebuck, Kroemer, and Thomson, 1975).

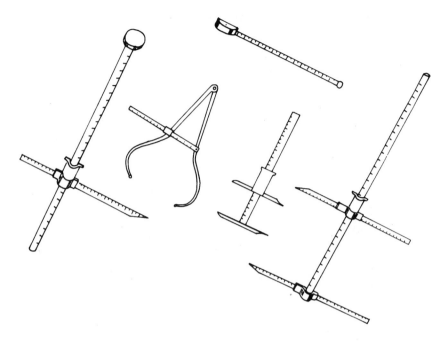

Figure 1-7. Classical anthropometric instruments: anthropometer, straight and spreading calipers, tape.

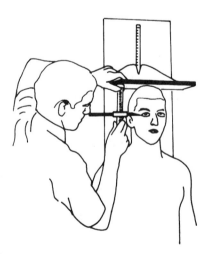

Figure 1-8. Typical use of the anthropometer with a sliding head board.

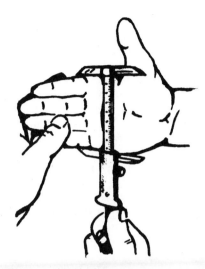

Figure 1-9. Measuring with a sliding caliper and using the anthropometer as a sliding rule.

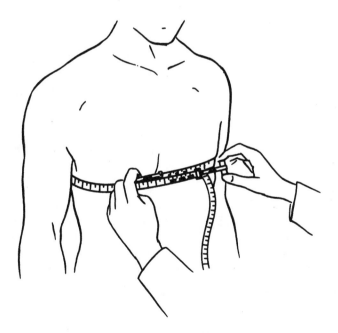

Figure 1-10. Circumference measurement with a tape.

New Measurement Methods

The classical anthropometric techniques rely on measurement instruments applied by the hand of the measurer to the body of the subject. Instruments and procedure are simple, relatively easy, and inexpensive. However, the process is somewhat clumsy and certainly time-consuming. Each measurement and tool must be selected in advance; and what was not measured in the test session remains unknown unless the subject is called back for more measuring. Another major disadvantage is that many of the classical dimensions are not related to each other in space. For example, as one looks at a subject from the side, stature, eye height, and shoulder height are located in different yet undefined frontal planes. Furthermore, certain parts of the body, such as the eyes, are very sensitive and cannot be touched.

Photographic methods overcome many of these disadvantages. They can instantly record all three-dimensional aspects of the human body. They allow storage of practically infinite numbers of measurements, which can be taken at one's convenience from the record. However, they also have drawbacks: equipment (particularly for data analyses) is expensive, a scale may be difficult to establish, parallax distortions must be eliminated, and landmarks cannot be palpated under the skin on the photograph.

It is for these and other reasons that, until the 1970s, photographic anthropometry was not commonly used, although systematic attempts had been made at least a century earlier. Since then, many improvements have been introduced by the use of several cameras, of mirrors, of film and videotape instead of still photography, of holography, and of various methods of stereophotometry.

A newer technique uses the laser as a distance-measuring technique. Either the laser moves around the body or the body moves in front of the laser. Coupled with a computer to store, sort, and reproduce the distance data, the human body surface can be described in minute detail, in three dimensions, with topographic techniques of great sensitivity. Yet points of interest below the surface, such as joints and landmark (e.g., the acromion) locations, do not appear in the data set unless "markers" are placed on the skin. A reliable and efficient procedure for converting the immense number of data points generated by laser technique, once developed and standardized, will allow the exact description of shapes of the body and its parts, and of changes in shape by motion, training, or aging. This will require means to convert the older anthropometric information to the new data format.

Obviously, anthropometric technology is very much in flux. Classical measuring techniques are likely to retain some importance because they are so simple and easily applied, even under adverse conditions, particularly in static or semi-static situations. However, high-technology anthropometric methods are gaining importance and will certainly continue to do so.

BODY TYPOLOGY

We can try to assess human body dimensions by using descriptors of body components and how they "fit" together. Our images of the beautiful body are affected by aesthetic codes, canons, and rules founded on often ancient (that is, Egyptian, Greek, and Roman) concepts of the human body. A more recent example is Leonardo de Vinci's drawing of the body within a frame of graduated circles and squares as a symbol of the well-proportioned body; it has been adopted, in simplified form, as the emblem of the *Human Factors and Ergonomics Society.*

Categorization of body builds into different types is called somatotyping, from the Greek *soma* for "body." In about 400 B.C., Hippocrates developed a scheme that included four body types, which were supposedly determined by their fluids. The moist type was dominated by black gall; the dry type was governed by yellow gall; the cold type was characterized by its slime, and the warm type was governed by blood. In *Körperbau und Charakter* (1921), the psychiatrist Ernst Kretschmer set forth a system of three body types intended to relate physical build to personality traits. Kretschmer's typology consisted of the asthenic, pyknic, and athletic body build. (The athletic type was to indicate character traits, not sportive performance capabilities.) In the 1940s, the anthropologist W.H. Sheldon established a system of three body types that was intended to describe (male) body proportions. Sheldon rated each person's appearance with respect to ecto-, endo-, and mesomorphic components (see Table 1-1). It is of interest to note that Sheldon's typology was originally based on intuitive assessment, not on actual body measurements: these were introduced into the system by his disciples.

These and other attempts at somatotyping (such as the Heath–Carter system) have not provided reliable predictors of attitudes nor of capabilities or limitations regarding human performance in technological systems. Hence, somatotyping is of little value for engineers or managers.

ANTHROPOMETRIC DATA SETS

The military has always had a particular interest in the body dimensions of soldiers for a variety of reasons, among them the necessity to provide uniforms, armor, and equipment that fit. Furthermore, armies have had surgeons and medical personnel who are in a position to perform body measurements on large samples "on command." Hence, anthropometric information about soldiers has a long history and is rather complete. The anthropometric data bank of the Armstrong Human Engineering Division, U.S. Air Force (Wright-Patterson AFB, OH 45433-7022) contains the data from a large number of military or civilian surveys from many nations, but the majority of the data consists of dimensions measured on U.S. Navy, Army, and Air Force personnel. The NASA *Anthropometric Sourcebook*

Table 1-1. Body typologies.

Descriptors	Stocky, Stout, Soft, Round	Strong, Muscular, Sturdy	Lean, Slender, Fragile
Kretschmer typology	pyknic	athletic	asthenic (leptosomic)
Sheldon typology	endomorphic	mesomorphic	ectomorphic

(1978) provides a good overview of the data available up to the mid-1970s. In 1986, Kennedy collated the data available on U.S. Air Force personnel. Among the most recent and comprehensive information is a report on data measured in 1988 on U.S. Army personnel (Gordon et al., 1989). Dimensions of the head, hand, and foot measured on military populations are apparently similar to those of civilians (NASA, 1978: Garrett and Kennedy, 1971; White, 1982) while other dimensions may reflect that soldiers are a select sample of the general population.

Normality

Fortunately, anthropometric data are usually distributed in a reasonably normal (Gaussian) distribution. Hence, regular parametric statistics apply in most cases. The most common statistical procedures are listed in Table 1-2.

Table 1-2.   Statistical formulas of particular use in anthropometry.

Measures of Central Tendency

Mean, Average	$m = \Sigma x/n$	(1st Moment)	(A-1)
Median	Middle value (of values in numerical order), 50th percentile.		
Mode	Most often found value		

Measures of Variability

Range	$x_{max} - x_{min}$		(A-2)
Standard Deviation	$S = (\text{Variance})^{1/2} = [\Sigma(x - m)^2/(n - 1)]^{1/2}$	(2nd Moment)	(A-3)
Coefficient of Variation	$CV = S/m$ Std Dev/mean		(A-4)
Standard Error of the Mean	$SE = S/n^{1/2}$		(A-5)
Two-Sided Confidence Limits for Mean	$m \pm k\ SE$ (see Table 1-8 for k)		(A-6)
Skewness	$\Sigma(x - m)^3/n$	(3rd Moment)	(A-7)
Peakedness	$\Sigma(x - m)^4/n$	(4th Moment)	(A-8)

Measures of Relationship between Two Variables x and y

Correlation Coefficient	$r = S_{xy}/(S_x S_y)^{1/2} =$ $\Sigma[(x - m)\,(y - \overline{y})]/[\Sigma\,(x - m)^2\,\Sigma(y - \overline{y}^2)]^{1/2}$	(A-9)
Regression	$y = ax + b$	(A-10)
	$a = rS_y/S_x$	(A-11)
	$SE_y = S_y(1 - r^2)^{1/2}$	(A-12)

The first and easiest check is on mean, median, and mode; if they concur, normality is likely. Another method is to calculate the average *m* first by using the complete range of data and then by leaving out the (say, 10) smallest and largest numbers; if both calculations end up with the same mean, normality is often the case. More formal calculations rely on the measures of symmetry and of peakedness contained in Table 1-2. Regarding skewness, a result of 0 indicates a symmetrical distribution, and a positive (negative) result points to skewness to the left (right). Regarding peakedness (kurtosis), a normal distribution is usually assumed when the numerical result is 3; if the result is larger (smaller) than 3, the distribution is peaked (flat). For further details, please refer to a statistics textbook.

Variability

Edmund Churchill, the eminent statistician in the field of engineering anthropometry during the second half of the 20th century, said (NASA, 1978, p. IX-5):

> A pioneer in the field of statistics, Sir Francis Galton [1822–1911] wrote years ago that "it is difficult to understand why statisticians commonly limit their interests to averages. Their souls seem as dull to the charm of variety as that of a native of one of our flat English counties whose retrospect of Switzerland was that, if its mountains would be thrown into its lakes, two nuisances could be got rid at once." Basic to virtually all design problems is the fact that mankind is far more like Switzerland than a flat English county, and that, whatever the charms of variety may be, we need statistics to quantify this variety.

To help assess the variability found in anthropometric data, the standard deviation is a very useful statistic. It describes the dispersion of a given set of anthropometric data, which reflects the (true) variability of the underlying data and the accuracy and reliability of the measuring technique applied. For example, if we compare the descriptive statistics of three distinct though related population samples and find that one shows a much larger coefficient of variation (standard deviation divided by the mean) of like dimensions than the others, we deduce from this result that either the sample with the large dispersion was, in fact, more variable than the other two samples or that there was suspicious variability in the measuring technique or in the data recording.

Correlations

Some body dimensions are closely related to each other; for example, eye height is very highly correlated with stature, but head length, waist circumference, and hip breadth are not. Table 1-3 shows selected correlation coefficients among body dimensions of U.S. Air Force personnel, male and female. (A more detailed table is contained in Appendix 1-A at the end of this chapter.)

The correlation coefficient provides concise information about the relationship between two or more sets of data. It is common practice in anthropometry (in fact, in all human engineering) to require a coefficient of at least 0.7 in order to base design decisions on this correlation. The reason for selecting this coefficient

Table 1-3. Selected correlation coefficients for anthropometric data on U.S. Air Force personnel: women above the diagonal, men below it (from NASA 1978).

	1	2	3	4	5	6	7	8	9	10
1. Age		.223	.048	−.023	.039	−.055	.091	−.072	.233	.287
2. Weight	.113		.533	.457	.497	.431	.481	.370	.835	.799
3. Stature	−.028	.515		.927	.914	.849	.801	.728	.334	.257
4. Chest height	−.028	.483	.949		.897	.862	.673	.731	.271	.183
5. Waist height	−.033	.422	.923	.930		.909	.607	.762	.308	.238
6. Crotch height	−.093	.359	.856	.866	.905		.467	.788	.264	.190
7. Sitting height	−.054	.457	.786	.681	.580	.453		.398	.312	.239
8. Popliteal height	−.102	.299	.841	.843	.883	.880	.485		.230	.172
9. Shoulder circumference	.091	.831	.318	.300	.261	.212	.291	.182		.810
10. Chest circumference	.259	.832	.240	.245	.203	.147	.171	.114	.822	
11. Waist circumference	.262	.856	.224	.212	.142	.132	.167	.068	.720	.804
12. Buttock circumference	.105	.922	.362	.334	.278	.217	.347	.149	.744	.766
13. Biacromial breadth	.003	.452	.378	.335	.339	.282	.349	.316	.555	.401
14. Waist breadth	.214	.852	.287	.260	.215	.195	.216	.133	.715	.801
15. Hip breadth	.105	.809	.414	.380	.342	.283	.376	.221	.632	.647
16. Head circumference	.110	.412	.294	.251	.233	.188	.287	.194	.327	.340
17. Head length	.054	.261	.249	.218	.208	.170	.244	.175	.201	.196
18. Head breadth	.122	.305	.133	.097	.089	.066	.132	.075	.245	.271
19. Face length	.119	.228	.275	.220	.226	.199	.253	.193	.162	.172
20. Face breadth	.233	.453	.190	.160	.142	.099	.185	.098	.401	.421

of determination as the minimum acceptable value lies in the convention that one should be able to explain at least 50% of the variance of the predicted value from the predictor variable; this requires r^2 to be at least 0.5, i.e., r at least 0.7075. (Note that r depends on sample size.)

This 0.7 convention is important for the development and use of regression equations, which express the average of one variable as a function of another (see Table 1-2 and the discussion section on "How to Get Missing Data"). If the use of only one predictor variable is not sufficient to establish an overall correlation coefficient between the predictor and a predicted value of at least 0.7, additional predictor variables may be taken into the equation until that minimal cutoff point is exceeded. Examples of such regression equations are in Table 1-4.

Body Proportions

In the past, some people have found it convenient to calculate ratios or proportional relationships of body dimensions, which are often called indices. Such index numbers are unjustified and misleading when there is an insufficient correlation between the two variables, as just discussed. In the 1960s, Drillis and Contini published a "stick man" figure that indicated ratios between various body segments and stature; this material was reprinted (with appropriate warnings and notes of caution) by Roebuck, Kroemer, and Thomson (1975) and by Chaffin and Andersson (1984). Unfortunately, some practitioners have used these ratios indiscriminately, even when the coefficients of correlation were well below the 0.7 convention. In Table 1-3, only chest height, waist height, crotch height (a measure

Table 1-3. (continued)

	11	12	13	14	15	16	17	18	19	20
1. Age	.234	.219	.149	.146	.194	.095	.118	.190	.189	.089
2. Weight	.824	.886	.495	.768	.770	.403	.304	.290	.264	.358
3. Stature	.279	.360	.456	.329	.348	.331	.318	.136	.267	.199
4. Chest height	.216	.289	.412	.266	.276	.284	.284	.085	.222	.162
5. Waist height	.238	.336	.409	.293	.318	.306	.297	.123	.225	.200
6. Crotch height	.221	.246	.380	.277	.225	.294	.280	.089	.205	.172
7. Sitting height	.236	.383	.384	.277	.379	.294	.275	.136	.248	.146
8. Popliteal height	.186	.201	.327	.249	.181	.235	.253	.087	.185	.189
9. Shoulder circumference	.775	.717	.581	.719	.606	.330	.248	.252	.217	.313
10. Chest circumference	.796	.674	.370	.706	.551	.273	.204	.255	.176	.273
11. Waist circumference		.722	.382	.886	.600	.281	.149	.267	.174	.310
12. Buttock circumference	.852		.396	.668	.893	.310	.214	.238	.180	.269
13. Biacromial breadth	.288	.355		.401	.361	.311	.239	.178	.266	.211
14. Waist breadth	.936	.849	.327		.576	.292	.168	.263	.182	.296
15. Hip breadth	.724	.895	.340	.760		.265	.183	.188	.155	.215
16. Head circumference	.309	.330	.251	.310	.288		.692	.430	.273	.299
17. Head length	.158	.195	.179	.164	.166	.779		.115	.311	.113
18. Head breadth	.265	.252	.188	.268	.227	.521	.058		.174	.497
19. Face length	.129	.186	.187	.151	.161	.315	.289	.148		.144
20. Face breadth	.412	.394	.278	.410	.364	.464	.131	.660	.206	

of leg length), sitting height, and popliteal height show correlation coefficients of more than 0.7 to stature; head length, shoulder breadth, hip breadth, and many other variables are not well correlated with stature. Hence, if we want to predict one dimension from another, we must carefully check whether the correlation between the two data sets is sufficiently high. In general, long bone (link) dimensions are reasonably well correlated with each other. Breadth and depth dimensions, respectively, correlate reasonably well within their groups but not with stature; the same is true for circumferences and for foot, hand, and head measures.

Anthropometric Data of Civilians

Measured anthropometric data on civilian populations are rather sparse, whereas the body dimensions of military personnel are well known. Since the military is a large, albeit biased (young and healthy) sample of the overall population, it appears logical to use the military data to infer dimensions for the general civilian population. There are, however, several problems involved. One is that soldiers may be so highly selected that they constitute a special sample that is not representative of the overall population. Another question is whether a sufficient number of dimensions were measured, both in the military and the general population, to allow an assessment of how well one data set can represent the other. McConville, Robinette, and Churchill (1981) addressed both problems. They selected the 1965 U.S. Health Examination Survey and compared it for males to the 1967 survey of the U.S. Air Force and the U.S. Army 1966 survey; the com-

Table 1-4. Regression equations *(all data in cm, except weight in pounds).

For Males:

Variable Predicted	Equations	Std. Error of Estimate	Resulting Correlation
Eye Height, standing	= (.9544 Stature) − 3.39	0.99	0.986
Acromion Height	= (.927 Stature) − (.233 Sitting Height) + (.042 Chest Circumference) − 8.1318	1.79	0.957
Elbow Height	= (.879 Stature) − (.629 Sitting Height) + (.674 Elbow Rest Height sitting) − 2.0578	1.29	0.960
Knuckle Height	= (.5536 Stature) − (.1982 Knee Height sitting) − 9.961	2.08	0.820
Hip Breadth	= (.688 Weight) + (.88 Sitting Height) − .0062	1.54	0.878
Functional Reach Forward	= (.102 Stature) + (.497 Knee Height sitting) + (.461 Buttock-Knee Length) + 10.4423	3.99	0.613
Vertical Reach, sitting	= (.8883 Stature) + (.2525 Sitting Height) − 39.70	3.13	0.842
Eye Height, sitting	= (.827 Sitting Height) + (.79 Elbow Rest Height sitting) + (.60 Buttock-Knee Length) − 1.769	1.09	0.933
Acromion Height, sitting	= (.202 Stature) + (.93 Sitting Height) + (.709 Elbow Rest Height sitting) − 1.2814	1.01	0.936
Forearm-Hand Length	= (.086 Stature) + (.298 Knee Height sitting) + (.234 Buttock-Knee Length) + 2.8683	1.32	0.821
Chest Depth	= (.263 Weight) − (.049 Sitting Height) + (.165 Chest Circumference) + 7.9929	1.17	0.811

For Females:

Variable Predicted	Equations	Std. Error of Estimate	Resulting Correlation
Eye Height, standing	= (.963 Stature) − 5.7101	1.07	0.984
Acromion Height	= (.957 Stature) − (.208 Sitting Height) + (.065 Waist Circumference) − 9.6449	1.47	0.967
Elbow Height	= (.6952 Stature) − 10.33	1.75	0.931
Knuckle Height	= (.4095 Stature) + (.2227 Sitting Height) − 14.371	2.19	0.833
Hip Breadth	= (1.338 Weight) − (.148 Popliteal Height) − (.100 Chest Circumference at Scye) + 32.4839	1.39	0.801
Overhead Reach, standing	= (1.066 Stature) − (.287 Elbow Rest Height) + (.505 Buttock-Knee Length) + 3.1681	4.28	0.866
Eye Height, sitting	= (.907 Sitting Height) − 3.7877	1.07	0.943
Elbow-Fingertip Length	= (.643 Popliteal Height) + (.175 Buttock-Knee Length) + 6.5701	1.19	0.855
Chest Circum. at Nipple	= (.8381 Chest Circumference at Scye) + (.0861 Weight) + 5.727	2.85	0.896
Chest Depth	= (.067 Stature) + (.613 Weight) + (.147 Chest Circumference at Scye) + 13.9110	1.34	0.766

*Modified from K. M. Robinette's personal communication of 10/22/81.

parison data for females came from the 1968 U.S. Air Force and the 1977 U.S. Army surveys. The underlying assumption for the comparisons was that, if good height and weight matches can be achieved between civilian and military individuals, then the means and standard deviations of other dimensions (measured in either survey) should be well matched also. This, of course, is a debatable assumption.

The procedure used was to match individuals from the civilian and military surveys on the basis of stature and weight (with matching intervals of ±1 in. and ±5 lb). Thus, a new military sample was created that represented the civilian sample in height and weight. From the "new" (matched) military sample, dimensions other than height and weight were selected and compared to the corresponding data measured in the civilian surveys.

For the males, an excellent fit was achieved: 99% of all civilians could be matched with at least one soldier. The mean differences of the samples in stature and weight were negligible. Comparison of the measures of six linear dimensions in both the military and civilian surveys provided similar good matches in means and standard deviations.

Use of regression equations to calculate one body dimension from others (height and weight) resulted in averages of the six civilian dimensions predicted from the military set that were very close to the results achieved in matching. However, standard deviations predicted from regression equations turned out to be considerably larger than those obtained in the matched pair procedure.

The same procedures were used in comparing female civilian and military data. About 94% of the civilian data could be matched with military individuals, with good fit in stature but poorer matches in weight. (The unmatched individuals were very short and very heavy civilians with weights between 90 and 135 kg. For these women, no military matches were found.)

Hence, with proper caution and insight, we can use military anthropometric data to approximate size data for the general population. Dimensions of the head, hand, and foot are virtually the same in military and civilian populations.

FUNCTIONAL ANTHROPOMETRY

Classical anthropometric data provide information on static dimensions of the human body in standard postures. However, these data do not describe functional performance capabilities, such as reach capabilities. These are traditionally measured with either the tip of a finger just touching an object (thumb-tip reach) or with the tips of several fingers enclosing a small object or with the whole hand grasping an object (grip or grasp reach). Figure 1-11 shows hand-object couplings. The reach measurement anatomically originates at the shoulder joint and encompasses an envelope around this joint with the arm extended where possible or bent in elbow and wrist when needed.

Although the shoulder joint as reference point makes some anatomical sense, this is not a practical origin for the data. Therefore, most reach studies have employed devices that define, by their construction, the reference point for the measures. Some data on reach envelopes employ the seat reference point (SRP) as

1.   Digit Touch:
   One digit touches an object without holding it.

2.   Palm Touch:
   Some part of the inner surface of the hand touches the object
   without holding it.

3.   Finger Palmar Grip (Hook Grip):
   One finger or several fingers hook(s) onto a ridge or handle.
   This type of finger action is used where thumb counterforce
   is not needed.

4.   Thumb-Fingertip Grip (Tip Pinch):
   The thumb tip opposes one fingertip.

5.   Thumb-Finger Palmar Grip (Pad Pinch):
   Thumb pad opposes the palmar pad of one finger or the pads
   of several fingers near the tips. This grip evolves easily from
   coupling #4.

6.   Thumb-Forefinger Side Grip (Lateral Grip or Side Pinch):
   Thumb opposes the radial side of the forefinger at its
   middle phalanx.

7.   Thumb–Two-Finger Grip (Writing Grip):
   Thumb and two fingers (often forefinger and index finger)
   oppose each other at or near the tips.

8.   Thumb-Fingertips Enclosure (Disk Grip):
   Thumb pad and the pads of three or four fingers oppose each
   other near the tips (object grasped does not touch the palm).
   This grip evolves easily from coupling #7.

9.   Finger-Palm Enclosure (Enclosure):
   Most, or all, of the inner surface of the hand is in contact with
   the object while enclosing it.

10.   Grasp (Power Grasp):
    The total inner hand surface is grasping the (often cylindrical)
    handle, which runs parallel to the knuckles and generally
    protrudes from one side or both sides of the hand.

Figure 1-11. Various couplings between hand and object, ranging from a mere *touch*
through *grip* and *enclosure* to the powerful *grasp*. (Based on the 1986 taxonomy
described in Coupling the Hand with the Handle: An Improved Notation of Touch, Grip,
and Grasp by K.H.E. Kroemer, *Human Factors* 28:337–339.)

origin, which is the point in the medial plane at which the surfaces of the seat back and of the seat pan meet. Figure 1-12 shows such a method to measure the reach envelope. However, many forward reach data are measured from a wall against which the subject leans.

Another topic of applied anthropometry is that of space needs and workplace dimensions for the body in common working postures. Since it is difficult to define working postures, which vary from task to task, very few data have actually been measured. Figure 1-13 shows some workplace information for male military personnel. For other populations and for specific work tasks, data can be derived from standard anthropometry or must be specifically measured to provide the information needed in a given case.

CAUSES OF VARIABILITY OF ANTHROPOMETRIC DATA

Causes and symptoms of variability in anthropometric data can be divided into four groups: interindividual variations, intraindividual variations, secular changes, and poor data.

Interindividual Variations

Interindividual variations are a result of DNA characteristics, of which about 10^9 possible chromosome combinations exist. An individual's genetic endorsement determines his/her cellular composition (genotype) and biologically measurable characteristics (phenotype). In addition, the individual's body size is influenced by the environment, that is, by altitude, temperature, sunlight and, perhaps, soil type. Obviously, amount and kind of nutrition also have direct effects on body size.

A variety of "recommended" body weight tables exist, often subdivided for gender, age, and body build. Among the best-known tables, which are used by

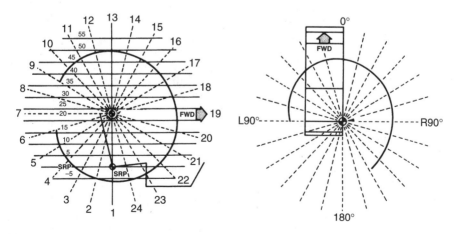

Figure 1-12. Example of the reach envelope measured from the seat reference point SRP in different horizontal planes: shown is the side view (left: height levels in inches) and the top view (right) (from Roebuck, Kroemer, and Thomson, 1975).

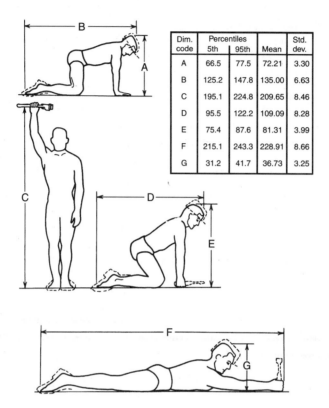

| Dim. | Percentiles | | | Std. |
code	5th	95th	Mean	dev.
A	66.5	77.5	72.21	3.30
B	125.2	147.8	135.00	6.63
C	195.1	224.8	209.65	8.46
D	95.5	122.2	109.09	8.28
E	75.4	87.6	81.31	3.99
F	215.1	243.3	228.91	8.66
G	31.2	41.7	36.73	3.25

Figure 1-13. Space needs in cm of male military personnel for nonstandard working postures (from Roebuck, Kroemer, and Thomson, 1975).

many physicians, are those prepared in 1942 and since updated by the Metropolitan Life Insurance Company. They all rely on the assumption that desirable weights can be gleaned from population statistics arranged for the dependent variable mortality from specific diseases (heart, diabetes, cancer, stroke). The underlying idea was that the weights of 20- to 25-year-old persons were "ideal" and should be maintained throughout life, with modifications for gender and body build. Andres (1984) questioned the validity of the population statistics on a number of anthropometric issues: that persons seeking insurance may not represent a random sample of the general population; that heights and weights were not carefully measured; that height and weight measurements by themselves are poor estimates of body obesity; and that no measurements of body build were ever made. Body weight is not consistently correlated with stature: it may be about 0.6 within highly selected groups (e.g., flight attendants), about 0.5 among soldiers, and much lower within the general population. Nevertheless, many people orient themselves by the presumed "desirable" weight-height relations, some of which are shown in Table 1-5.

Table 1-5. Height-Weight recommendations for "desirable" body weights in lb by Metropolitan Life and Gerontology Research Center versus obesity criteria the National Institutes of Health (excerpts).

Height	Metropolitan Life (1983) "medium frame"		Gerontology Research Center (1985)[a]					Obesity NIH (1985)	
	Men	Women	20–29 yr	30–39 yr	40–49 yr	50–59 yr	60–69 yr	Men	Women
4'10"	—	100–131	84–111	92–119	99–127	107–135	115–142	—	137
5'0"	—	103–137	90–119	98–127	106–135	114–143	123–152	—	143
5'2"	128–138	108–144	96–127	105–136	113–144	122–153	131–163	160	150
5'4"	129–155	114–152	102–135	112–145	121–154	130–163	140–173	164	157
5'6"	133–163	120–160	109–144	119–154	129–164	138–174	148–184	172	164
5'8"	137–171	126–167	116–153	126–163	137–174	147–184	158–196	179	172
5'10"	141–179	132–173	122–162	134–173	145–184	156–195	167–207	186	179
6'0"	147–187	—	129–171	141–183	153–195	165–207	177–219	194	—
6'2"	153–197	—	137–181	149–194	162–206	174–219	187–232	203	—
6'4"	—	—	144–191	157–205	171–218	184–231	197–244	—	—

[a]Same values for women and men.

The population of the United States (and of some other countries) is a composite of many ethnic origins. For example, about 100 million U.S. citizens say they have either English or German roots, while approximately 40 million claim Irish ancestry; nearly 21 million are of African descent; about 8 million stem from Mexico, and more than 2.5 million are of East Asian origin. Table 1-6 contains more details. There are statistically significant anthropometric differences between groups of various ethnic origins and, while these differences may be of practical importance for the design and use of certain items in defined localities, on a nationwide scale these differences are of no great magnitude and of little practical importance (see, e.g., NASA, 1978). The variations in body size within groups are usually much more striking than the average differences between groups. This is also true for differences in body sizes among various professions; on a large scale, white-collar and blue-collar groups are not much different anthropometrically.

Handedness needs to be carefully defined: for example, one may prefer the left hand for writing but the right hand for hammering. The overall estimate is that about 10% of all Americans are left-handed. This may be of some interest to, say, the manufacturers of special hand tools but otherwise does not imply large differences in arm lengths, or arm circumferences, as reflected in nationwide anthropometric tables.

Table 1-6. Ancestry claimed by percentage of U.S. citizens (Source: U.S. Census Bureau Poll 1990).

English	13.1	Chinese	0.6
German	23.3	Filipino	0.6
Irish	15.6	Japanese	0.4
Afro-American	9.6	French Canadian	0.9
French	4.1	Slovak	0.8
Italian	5.9	Lithuanian	0.3
Scottish	4.5	Ukranian	0.3
Polish	3.8	Finnish	0.3
Mexican	4.7	Cuban	0.3
American Indian	3.5	Canadian	0.2
Dutch	2.5	Korean	0.3
Swedish	1.9	Belgian	0.2
Norwegian	1.6	Asian Indian	0.2
Russian	1.2	Lebanese	0.2
Spanish-Hispanic	0.8	Jamaican	0.2
Czech	0.5	Croatian	0.2
Hungarian	0.6	Vietnamese	0.2
Welsh	0.8	Dominican	0.2
Danish	0.7		
Puerto Rican	0.8		
Portuguese	0.5	All others 0.1 or less	
Swiss	0.4		
Greek	0.4		
Austrian	0.3		

Listed according to the claims reported in the 1980 poll.

Intraindividual Variations

Among the intraindividual variables, the effects of aging on anthropometry are rather obvious. During the growing years, stature, weight, and other body dimensions increase, which then become relatively stable in early adulthood. With increasing age, certain dimensions begin to decrease (such as body height) while circumferences and the external diameters of bones usually increase. Table 1-7 lists, in approximate numbers, changes in stature with age.

Other examples of intraindividual variations are variations in weight or circumferences associated, for instance, with changes in nutritional and physical activities. (In fact, part of the foregoing discussion on desirable weights concerned intraindividual aspects.) Variations in stature during the day are also a case of changes within a person. Immediately after rising from bedrest in the morning, we may be several centimeters taller than after a full day on our feet. This is due mainly to the loss of body fluids from the intervertebral disks as a result of compressive forces generated by gravity and body activities.

Secular Variations

The term *secular* indicates time-related events. When looking at medieval armor displayed in a museum, we cannot help but think how small the soldiers must have been centuries ago. In fact, there is some factual and much anecdotal evidence that people are nowadays, on the average, larger than their ancestors. However, "hard" anthropometric information on this development is only available for approximately the last hundred years. From the mid-19th century on, anthropometric surveys were conducted in a reasonably consistent manner and on sufficiently large samples to yield reliable results. Figure 1-14 provides information on stature for a

Table 1-7.   Approximate changes in stature with age.

Age (in years)	Change (in cm)	
	Females	Males
1 to 5*	+36	+36
5 to 10	+28	+27
10 to 15	+22	+30
15 to 20	+1	+6
20 to 35**	0	0
35 to 40	−1	0
40 to 50	−1	−1
50 to 60	−1	−1
60 to 70	−1	−1
70 to 80	−1	−1
80 to 90	−1	−1

*Average stature at age 1: females 74 cm, males 75 cm.

**Average maximal stature: females 161 cm, males 174 cm.

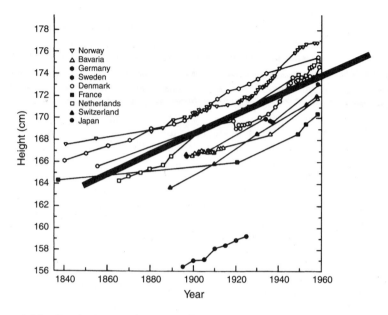

Figure 1-14. Secular increase in stature of young European and Japanese males. Heavy line shows the (estimated) apparent trend. (Adapted from NASA, 1978.)

variety of civilian samples: the overall trend is apparent. Similarly, Figure 1-15 presents information on military data measured in the United States. It is surprising that the military data show virtually no change between the Civil War and World War I but a pronounced increase thereafter. Why stature remained seemingly stable for nearly 60 years and then increased rapidly is open to speculation. Perhaps the recruitment of soldiers from the general population was different during the Civil War from the screening process during World War I; or perhaps the general population had a massive influx of shorter immigrants; or perhaps measurement techniques have been systematically different. (This brief discussion points out some of the difficulties in comparing data that are disjointed in both time and collection technique.) Nevertheless, the increase observed in stature during the last 50 years is apparently "real." Data from virtually all major surveys in the United States and Europe indicate an increase in stature of about 1 cm per decade during the 20th century. Weight increases were even more dramatic, in the neighborhood of 2 kg for every 10 years.

It is interesting to speculate about the reasons for these increases and to forecast the future development (Roebuck, 1995). One generally accepted explanation is that improvements in living conditions, both hygienic and nutritional, have allowed people to achieve their genetically possible stature more readily than in earlier times. If this is true, we would expect that, nationwide, an "average maximal height" would be approached in the future asymptotically, provided that

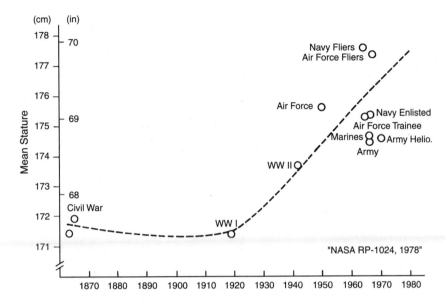

Figure 1-15. Trend of change in stature for U.S. soldiers (NASA, 1978).

improvement in living conditions applies to the whole population. Of course, such popular trends as weight consciousness and exercise (if generally practiced over long periods of time) could alter the development of body weight, one anthropometric characteristic.

Altogether, the secular developments of body dimensions are rather small and slow. Hence, for most engineers, the changes in body data should have few practical consequences for the design of tools, equipment, and workplaces since virtually none are designed to be used over many decades or even centuries. Most products have a relatively short design life, for which secular changes in anthropometry of the users have no appreciable importance.

Data Variations

The final—and undesirable—variation in anthropometric information may simply be due to sloppy sampling, measurement, recording, sorting, analysis, or reporting. All these faults may result in data collections that are "unusual" in central tendency (mean, median, mode) or variability compared to more valid data. Thus, if we encounter a report of body sizes that is at variance with related information, it is advisable to check carefully. A quick and valuable first test is to compare the "coefficient of variation" discussed later.

The Changing Population

Populations do not remain constant but change in age, health, strength, and so forth. Their composition changes as well; for example, the work force in the United States today has many more women in occupations that used to be dominated by males just a few decades ago. Occupations have changed drastically; for example, computer use has become widespread.

Computers and service industries are drawing people from traditional workshops and industries, and fewer workers are blue-collar in the traditional sense. Average life expectancy in the United States increased from 48 years in 1900 to 75 years in 1986, a difference of 27 years. In 2020, approximately 18% of all Americans will be 65 years and older. Thus, the convenient assumption of the broad-based "population pyramid," in which many young people support a few old people is no longer true. In 1971, the median age of the population was almost 28 years; in 1980, 30 years; in 1986, almost 32 years; it is expected to exceed 36 years by the year 2000. Currently, there are more Americans over 65 than there are teenagers.

The total fertility rate declined from 3.7 births per woman in 1960 to 2.5 in 1970 and to about 1.8 in 1985. While such a low birthrate would lead to a reduction in the total U.S. population within just a few decades, immigration keeps the number of U.S. citizens from shrinking. Estimates are that, in 100 years, the United States will have a population of about 300 million, of which about 16% will be of African, 16% of Hispanic, and 10% of Asian ancestry.

Americans are moving within their country. In 1980, for the first time in modern times, the majority of Americans lived in the southern and western states. The Sun Belt—prominently California, Florida, and Texas—showed greater population growth than the other regions in the United States; migration has reduced the population in the Snow Belt. Cities used to be magnets for the rural population; in the 1960s and 1970s, the flow was reversed but, in the 1980s, some metropolitan areas were again growing. Still, many people seem to move away from the very large cities to smaller, less crowded communities. These trends can change quickly, however, with economic developments, for example.

Such demographic changes require that anthropometric data describing the population in general and, more importantly, depicting certain groups of particular interest to the engineer, be carefully monitored so that their body dimensions can be considered properly in technical designs. For example: a computer workstation that fits a college group in North Dakota may not be of the correct size and adjustment range for, say, computer operators in southern Texas.

Designing to Fit the Body

While all humans have heads and trunks, arms and legs, these body parts come in various sizes and are assembled in different proportions. Measuring human bodies results in compilations of anthropometric data. Most body data appear, statistically speaking, in a normal (Gaussian) distribution. Such distribution of data can be described by using the statistical descriptors *mean* (same as *average*), *standard deviation,* and *range,* if the sample size is large enough (see below for more detail). Misunderstanding and misuse have led to the false idea that one could design for the average; yet the mean value is larger than half the data and smaller than the other half. Consequently, the average does not describe the ranges of different statures, arm lengths, or hip breadths. Furthermore, one is unlikely ever to encounter a person who displays mean values in several, many, or all dimensions. The "average person" is nothing but a statistical phantom.

Useful and correct steps in designing for clothing, tools, workstations, and equipment to fit the body are given by Kroemer, Kroemer, and Kroemer-Elbert (1994) as follows.

> **Step 1:** *Select those anthropometric measures that directly relate to defined design dimensions.* Examples: Hand length related to handle size; shoulder and hip breadth related to escape-hatch diameter; head length and breadth related to helmet size; eye height related to the heights of windows and displays; knee height and hip breadth related to the legroom in a console.
>
> **Step 2:** *For each of these pairings, determine whether the design must fit only one given percentile (minimal or maximal) of the body dimension, or a range along that body dimension.* Examples: The escape hatch must be big enough to accommodate the largest extreme value of shoulder breadth and hip breadth, considering clothing and equipment worn; the handle size of pliers is probably selected to fit a smallish hand; the legroom of a console must accommodate the tallest knee heights; the height of a seat should be adjustable to fit persons with short and with long lower legs. (How to use and calculate percentiles is explained in the following section.)
>
> **Step 3:** *Combine all selected design values in a careful drawing, mock-up, or computer model to ascertain that they are compatible.* Examples: The legroom clearance height needed for sitting persons with long lower legs may be very close to the height of the working surface determined from elbow height.
>
> **Step 4:** *Determine whether one design will fit all users.* If not, several sizes or adjustment must be provided to fit all users. Examples: One extra-large bed size fits all sleepers; gloves and shoes must come in different sizes; seat heights are adjustable.

Using Percentiles

Since most body dimensions are normally distributed, their distribution follows the well-known bell curve shown in Figure 1-16. Only a few persons are very short, or very tall, but many cluster around the center of the distribution (the mean or average). Figure 1-16 shows an approximate stature distribution of male Americans; only 2.5% are shorter than approximately 1,620 mm, and the other about 2.5% are taller than 1,880 mm. In other words, about 95% of all men are in the height range 1,620 to 1,880 mm because the 2.5th percentile value is at 1,620 mm and the 97.5th percentile is at 1,880 mm. The 50th percentile is at 1,750 mm. (In a normal distribution, mean, average, median, and mode coincide with the 50th percentile.) The standard deviation is 67 mm.

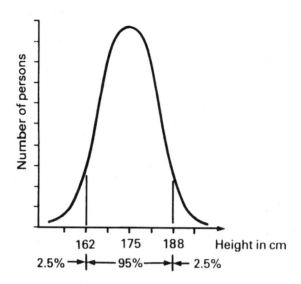

Figure 1-16. Frequency distribution of body height (stature) in Americans. About 95% of all males are between 162 and 188 cm tall; about 5% are either shorter or taller.

There are two ways to determine given percentile values. One is simply to take a distribution of data such as the one shown in Figure 1-16, and determine from the graph (measure, count, or estimate) critical percentile values. This works well whether the distribution is normal, skewed, binomial, or in any other form. Fortunately, most anthropometric data are normally distributed, which allows the second, even easier (and usually more exact) approach: to calculate percentile values. This involves the standard deviation s. If the distribution is flat (the data are widely scatttered), the value of s is larger than it is when the data cluster close to the mean.

To calculate the percentile value, we simply multiply the standard deviation s by a factor k, selected from Table 1-8. Then, we add the product to the mean m:

$$p = m + ks \qquad (1\text{-}1)$$

If the desired percentile is above the 50th percentile, the factor k has a positive sign and the product ks is added to the mean; if the p-value is below average, k is negative and the product ks is, in fact, subtracted from the mean m.

Examples:

1st percentile is at $m - ks$ $k = -2.33$ (see Table 1-8)
2nd percentile is at $m - ks$ $k = -0.205$
2.5th percentile is at $m - ks$ $k = -1.96$

Table 1-8. Percentile values and associated k factors. Any percentile value p can be calculated from the mean m and the standard deviation s (normal distribution assumed) by $p = m + ks$.

Below mean				Above mean			
Percentile	Factor k	Percentile	Factor k	Percentile	Factor k	Percentile	Factor k
0.001	−4.25	**25**	**−0.67**	**50**	**0**	76	0.71
0.01	−3.72	26	−0.64	51	0.03	77	0.74
0.1	−3.09	27	−0.61	52	0.05	78	0.77
0.5	−2.58	28	−0.58	53	0.08	79	0.81
1	−2.33	29	−0.55	54	0.10	**80**	**0.84**
2	−2.05	**30**	**−0.52**	**55**	**0.13**	81	0.88
2.5	−1.96	31	−0.50	56	0.15	82	0.92
3	−1.88	32	−0.47	57	0.18	83	0.95
4	−1.75	33	−0.44	58	0.20	84	0.99
5	**−1.64**	34	−0.41	59	0.23	**85**	**1.04**
6	−1.55	**35**	**−0.39**	**60**	**0.25**	86	1.08
7	−1.48	36	−0.36	61	0.28	87	1.13
8	−1.41	37	−0.33	62	0.31	88	1.18
9	−1.34	38	−0.31	63	0.33	89	1.23
10	**−1.28**	39	−0.28	64	0.36	**90**	**1.28**
11	−1.23	**40**	**−0.25**	**65**	**0.39**	91	1.34
12	−1.18	41	−0.23	66	0.41	92	1.41
13	−1.13	42	−0.20	67	0.44	93	1.48
14	−1.08	43	−0.18	68	0.47	94	1.55
15	**−1.04**	44	−0.15	69	0.50	**95**	**1.64**
16	−0.99	**45**	**−0.13**	**70**	**0.52**	96	1.75
17	−0.95	46	−0.10	71	0.55	97	1.88
18	−0.92	47	−0.08	72	0.58	98	2.05
19	−0.88	48	−0.05	73	0.61	99	2.33
20	**−0.84**	49	−0.03	74	0.64	99.5	2.58
21	−0.81	**50**	**0**	**75**	**0.67**	99.9	3.09
22	−0.77					99.99	3.72
23	−0.74					99.999	4.26
24	−0.71						

5th percentile is at $m - ks$ $k = -1.64$
10th percentile is at $m - ks$ $k = -1.28$
50th percentile is at m $k = 0$
90th percentile is at $m + ks$ $k = 1.28$
95th percentile is at $m + ks$ $k = 1.64$

Percentiles serve the designer in several ways. First, they help to establish the portion of a user population that will be included in (or excluded from) a specific design solution. For example, a certain product may need to fit everybody who is taller than the 5th percentile and smaller than the 60th percentile in hand size or arm reach. Thus, only the 5% having values smaller than the 5th percentile and the 40% having values larger than the 60th percentile will not be fitted while 55% (60% − 5%) of all users will be accommodated.

Second, percentiles are easily used to select subjects for fit tests. For example, if the product needs to be tested, persons having 5th or 60th percentile values in the critical dimensions can be employed for use tests.

Third, any body dimension, design value, or score of a subject can be exactly located. For example, a certain foot length can be described as a given percentile value of that dimension, or a certain seat height can be described as fitting a certain percentile value of lower leg length (e.g., popliteal height), or a test score can be described as falling at a certain percentile value.

Fourth, the use of percentiles helps in the selection of persons to use a given product. For example, if a cockpit of an airplane is designed to fit the 5th to 95th percentiles, cockpit crews can be selected whose body measures are at or between the 5th and 95th percentiles in the critical design dimensions.

To determine a single (distinct) percentile point:
1. Select the desired percentile value.
2. Determine the associated k value from Table 1-8.
3. Calculate the p value from $p = m + ks$. (Note that k, and hence the product, may be negative.)

To determine a range:
1. (a) Select the upper percentile p_{max}.
   (b) Find the related k_{max} value in Table 1-8.
   (c) Calculate the upper percentile value $p_{max} = m + k_{max} s$.
2. (a) Select lower percentile p_{min}. (Note that the two percentile values need not be at the same distance from the 50th p; i.e., the range does not have to be symmetrical to the mean.)
   (b) Find related k_{min} value in Table 1-8.
   (c) Calculate lower percentile value $p_{min} = m + k_{min} s$.
3. Determine range $R = p_{max} - p_{min}$.

To determine tariffs:
A distribution of body dimensions is often divided into certain sections, such as in establishing clothing tariffs. An example is the use of neck circumference to establish selected collar sizes for men's shirts. The first step is to establish the ranges (see the preceding paragraph), which shall be covered by the tariff sections. The second step is to associate other body dimensions with the primary one, for example, chest circumference or sleeve length with collar (neck) circumference. This can become a rather complex procedure because the combination of body dimensions (and their derived equipment dimensions) depends on correlations among these dimensions, as already discussed in the section on Correlations. For more information, see Roebuck, Kroemer, and Thompson (1975); McConville's Chapter VIII in NASA (1978); and Roebuck (1995).

Body Postures

To unify measurements, the body is put into standard postures as follows.

> *Standing:* The instruction is to stand erect; heels together; rears of heels, buttocks, and shoulders touching a vertical wall; head erect; look straight ahead; arms hang straight down (or upper arms hang, forearms are horizontal and extended forward); fingers extended.
>
> *Sitting:* on a plane, horizontal, hard surface adjusted in height so that the thighs are horizontal. The instruction: Sit with lower legs vertical, feet flat on the floor; trunk and head erect; look straight ahead; arms hang straight down (or upper arms hang, forearms horizontal and extended forward); fingers extended.

The *head* (including the neck) is held erect (or "upright") when, *in the front view,* the pupils are aligned horizontally and, *in the side view;* the ear-eye line is angled about $P = 15$ deg above the horizon (see Fig. 1-17).

People do not stand or sit in these postures naturally. Thus, the dimensions taken on the body in the standardized postures must be converted to reflect real postures. Examples were shown in Figure 1-13. The postures assumed at work, or at leisure, can be greatly varied. Therefore, it is impossible to give "conversion factors" that apply to all conditions. The designer has to estimate the corrections that reflect the anticipated postures. Some general guidelines are presented in Table 1-9.

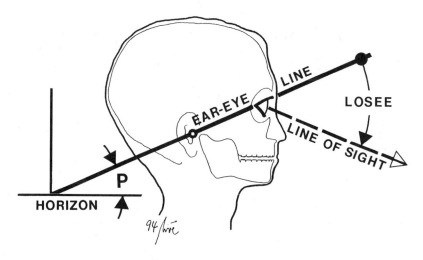

Figure 1-17. The ear-eye line serves as a reference to describe head posture (angle *P*) and the line-of-sight angle (LOSEE). The EE line passes through the landmarks ear hole and juncture of the eyelids.

Table 1-9. Guidelines for the conversion of standard measuring postures to functional work postures.

Slumped standing or sitting:	Deduct 5–10% from appropriate height measurements.
Relaxed trunk:	Add 5–10% to trunk circumference and depths.
Wearing shoes:	Add approximately 25 mm to standing and sitting heights; more for "high heels."
Wearing light clothing:	Add about 5% to appropriate dimensions.
Wearing heavy clothing:	Add 15% to appropriate dimensions. (Note that mobility may be strongly reduced by heavy clothing.)
Extended reaches:	Add 10% or more for strong motions of the trunk.
Use of hand tools:	Center of handle is at about 40% hand length, measured from the wrist.
Forward bent head (and neck) posture:	Ear-eye line close to horizontal.
Comfortable seat height:	Add or subtract up to 10% to or from popliteal height.

AVAILABLE BODY SIZE DATA

The most recent information on body sizes of North American adults, females and males, is presented in Table 1-10a. These data were derived from measurements on U.S. Army personnel taken in 1988. In spite of the military sampling bias due to selection, these data are the best available for the total North American adult population. The main reservation is with respect to body weight, which is more variable in the civilian population than in the military. Head, foot, and hand sizes should not differ appreciably between soldiers and civilians.

Note that, for each body dimension, the data are presented for the 5th and 95th percentiles, with the female data to the right and the male data to the left. The mean (50th percentile) is also given, as is the standard deviation. This allows the calculation of values other than 5th and 95th percentiles by using the multiplication factors in Table 1-8. Figure 1-18 illustrates the body dimensions listed in Table 1-10. Tables 1-10b and 1-10c list comparable body dimensions for German and Japanese adults.

Most populations on earth have not been measured thoroughly and completely. Table 1-11 presents an overview of available recent anthropometric data on national, ethnic, and geographic populations. It is unfortunate that, in most cases, few people were measured; the small number n makes it unlikely that the statistical descriptors mean m and standard deviation s truly represent the underlying large population. Table 1-12 contains general estimates for main regions of the earth. Given the paucity of existing data in 1990 and even today, it is not surprising that the estimates must be applied with great caution. Note, for example, that the averages estimated for North Americans are several centimeters higher than actually measured.

Table 1-10a. Anthropometric measured data in millimeters of U.S. adults, 19 to 60 years of age, according to Gordon et al. (1989). The reference numbers of the dimensions are shown in Figure 1-18.

Dimension	*Men*				*Women*			
	5th percentile	Mean	95th percentile	SD	5th percentile	Mean	95th percentile	SD
1. Stature [99]	1647	1756	1867	67	1528	1629	1737	64
2. Eye height, standing [D19]	1528	1634	1743	66	1415	1516	1621	63
3. Shoulder height (acromion), standing [2]	1342	1443	1546	62	1241	1334	1432	58
4. Elbow height, standing [D16]	995	1073	1153	48	926	998	1074	45
5. Hip height (trochanter) [107]	853	928	1009	48	789	862	938	45
6. Knuckle height, standing	na	na	na	na	na	na	na	na
7. Fingertip height, standing [D13]	591	653	716	40	551	610	670	36
8. Sitting height [93]	855	914	972	36	795	852	910	35
9. Sitting eye height [49]	735	792	848	34	685	739	794	33
10. Sitting shoulder height (acromion) [3]	549	598	646	30	509	556	604	29
11. Sitting elbow height [48]	184	231	274	27	176	221	264	27
12. Sitting thigh height (clearance) [104]	149	168	190	13	140	160	180	12
13. Sitting knee height [73]	514	559	606	28	474	515	560	26
14. Sitting popliteal height [86]	395	434	476	25	351	389	429	24
15. Shoulder-elbow length [91]	340	369	399	18	308	336	365	17
16. Elbow-fingertip length [54]	448	484	524	23	406	443	483	23
17. Overhead grip reach, sitting [D45]	1221	1310	1401	55	1127	1212	1296	51
18. Overhead grip reach, standing [D42]	1958	2107	2260	92	1808	1947	2094	87
19. Forward grip reach [D21]	693	751	813	37	632	686	744	34
20. Arm length, vertical [D3]	729	790	856	39	662	724	788	38
21. Downward grip reach [D43]	612	666	722	33	557	700	664	33
22. Chest depth [36]	210	243	280	22	209	239	279	21
23. Abdominal depth, sitting [1]	199	236	291	28	185	219	271	26
24. Buttock-knee depth, sitting [26]	569	616	667	30	542	589	640	30
25. Buttock-popliteal depth, sitting [27]	458	500	546	27	440	482	528	27
26. Shoulder breadth (biacromial) [10]	367	397	426	18	333	363	391	17
27. Shoulder breadth (bideltoid) [12]	450	492	535	26	397	433	472	23
28. Hip breadth, sitting [66]	329	367	412	25	343	385	432	27
29. Span [98]	1693	1823	1960	82	1542	1672	1809	81
30. Elbow span	na	na	na	na	na	na	na	na
31. Head length [62]	185	197	209	7	176	187	198	6
32. Head breadth [60]	143	152	161	5	137	144	153	5
33. Hand length [59]	179	194	211	10	165	181	197	10
34. Hand breadth [57]	84	90	98	4	73	79	86	4
35. Foot length [51]	249	270	292	13	224	244	265	12
36. Foot breadth [50]	92	101	110	5	82	90	98	5
37. Weight (kg), estimated by Kroemer	58	78	99	13	39	62	85	14

Numerals in brackets as used by Gordon et al. (1989)

Table 1-10b. Anthropometric measured data in millimeters of East German adults, 18 to 59 years of age, according to Fluegel, Greil and Sommer (1986). The reference numbers of the dimensions are shown in Figure 1-18.

	Men				*Women*			
Dimension	5th percentile	Mean	95th percentile	SD	5th percentile	Mean	95th percentile	SD
1. Stature	1607	1715	1825	66	1514	1608	1707	59
2. Eye height, standing	1498	1601	1705	64	1415	1504	1597	57
3. Shoulder height (acromion), standing	1320	1414	1512	60	1232	1319	1403	53
4. Elbow height, standing	na	na	na	na	na	na	na	na
5. Hip height (trochanter)	na	na	na	na	na	na	na	na
6. Knuckle height, standing	682	748	819	42	643	703	764	37
7. Fingertip height, standing	588	652	717	39	557	616	672	35
8. Sitting height	846	903	958	34	804	854	905	31
9. Sitting eye height	719	775	831	34	684	733	782	30
10. Sitting shoulder height (acromion)	552	601	650	31	517	562	609	29
11. Sitting elbow height	198	244	293	29	190	234	282	28
12. Sitting thigh height (clearance)	126	151	176	15	125	148	175	15
13. Sitting knee height	490	531	575	27	458	497	538	24
14. Sitting popliteal height	410	452	496	26	380	416	455	23
15. Shoulder-elbow length	na	na	na	na	na	na	na	na
16. Elbow-fingertip length	432	465	500	20	394	425	556	19
17. Overhead grip reach, sitting	na	na	na	na	na	na	na	na
18. Overhead grip reach, standing	1975	2121	2267	89	1843	1973	2103	79
19. Forward grip reach	704	763	824	37	650	706	767	35
20. Arm length, vertical	704	762	820	35	650	703	758	33
21. Downward grip reach	na	na	na	na	na	na	na	na
22. Chest depth	na	na	na	na	na	na	na	na
23. Abdominal depth, sitting	na	na	na	na	na	na	na	na
24. Buttock-knee depth, sitting	560	603	648	27	541	585	630	27
25. Buttock-popliteal depth, sitting	444	486	527	25	437	479	521	26
26. Shoulder breadth (biacromial)	365	399	430	20	336	365	393	17
27. Shoulder breadth (bideltoid)	432	471	510	24	393	437	481	27
28. Hip breadth, sitting	334	369	406	22	346	401	460	35
29. Span	1640	1760	1885	75	1503	1616	1735	70
30. Elbow span	833	895	911	39	757	817	881	38
31. Head length	179	190	201	7	170	181	191	6
32. Head breadth	148	158	168	6	141	151	160	6
33. Hand length	174	189	205	9	161	174	189	9
34. Hand breadth	81	88	96	5	71	78	85	4
35. Foot length	243	264	285	13	222	241	260	12
36. Foot breadth	91	102	113	6	83	93	104	6
37. Weight (kg)	na	na	na	na	na	na	na	na

Table 1-10c. Anthropometric measured data in millimeters of Japanese adults, 18 to 30 years of age, according to Kagimoto (1990). The reference numbers of the dimensions are shown in Figure 1-18.

	Men				*Women*			
Dimension	5th percentile	Mean	95th percentile	SD	5th percentile	Mean	95th percentile	SD
1. Stature	1599	1688	1777	55	1510	1584	1671	50
2. Eye height, standing	1489	1577	1664	53	1382	1460	1541	49
3. Shoulder height (acromion), standing	1291	1370	1454	50	1208	1279	1367	48
4. Elbow height, standing	970	1035	1098	39	909	967	1028	37
5. Hip height (trochanter)	775	834	899	38	730	787	847	35
6. Knuckle height, standing	na	na	na	na	na	na	na	na
7. Fingertip height, standing	600	644	694	30	563	608	652	27
8. Sitting height	859	910	958	30	810	855	902	28
9. Sitting eye height	741	790	837	29	692	733	778	27
10. Sitting shoulder height (acromion)	549	591	633	26	513	551	588	24
11. Sitting elbow height	216	254	292	23	202	236	269	20
12. Sitting thigh height (clearance)	138	156	176	12	130	143	162	10
13. Sitting knee height	475	509	545	22	442	475	508	20
14. Sitting popliteal height	371	402	434	19	345	372	402	17
15. Shoulder-elbow length	307	337	366	18	289	315	339	15
16. Elbow-fingertip length	418	448	479	18	390	416	445	17
17. Overhead grip reach, sitting	na	na	na	na	na	na	na	na
18. Overhead grip reach, standing	na	na	na	na	na	na	na	na
19. Forward grip reach	na	na	na	na	na	na	na	na
20. Arm length, vertical	na	na	na	na	na	na	na	na
21. Downward grip reach	na	na	na	na	na	na	na	na
22. Chest depth	190	217	246	18	190	215	250	19
23. Abdominal depth, sitting	179	208	245	20	161	188	218	17
24. Buttock-knee depth, sitting	530	567	604	23	511	550	586	22
25. Buttock-popliteal depth, sitting	na	na	na	na	na	na	na	na
26. Shoulder breadth (biacromial)	368	395	423	17	346	367	391	14
27. Shoulder breadth (bideltoid)	na	na	na	na	na	na	na	na
28. Hip breadth, sitting	318	349	380	19	331	358	386	17
29. Span	1591	1690	1795	63	1483	1579	1693	62
30. Elbow span	na	na	na	na	na	na	na	na
31. Head length	178	190	203	7	168	177	187	6
32. Head breadth	152	161	171	6	143	151	160	6
33. Hand length	na	na	na	na	na	na	na	na
34. Hand breadth	79	85	91	4	70	75	81	3
35. Foot length	234	251	269	11	217	232	246	9
36. Foot breadth	97	104	111	5	89	96	103	4
37. Weight (kg)	54	66	80	8	45	54	65	6

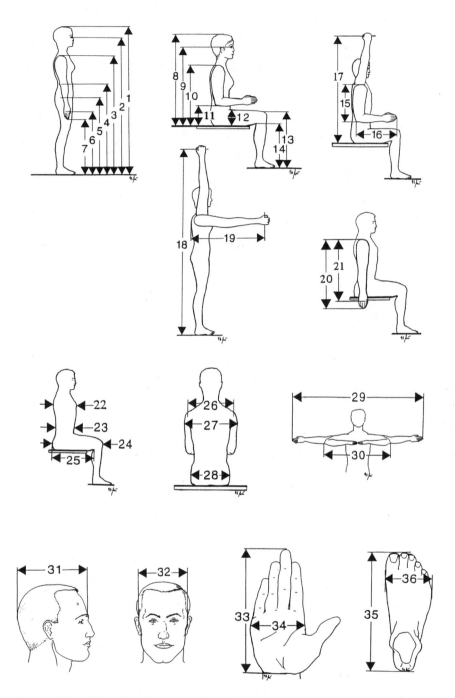

Figure 1-18. Illustration of measured body dimensions.

Table 1-11. Recent anthropometric data on national and ethnic populations: averages (and standard deviations) in millimeters; weight in kilograms.

	Sample size N	Stature	Sitting height	Knee height, sitting	Weight
Algerian females (1990)	666	1576 (56)	795 (50)	487 (36)	61.3 (12.9)
Brazilian males (1988)	3076	1699 (67)	—	—	—
Chinese females (Singapore) (1988)	46	1598 (58)	855 (31)	—	—
Cantonese males (1990)	41	1720 (63)	—	—	60.0 (6.2)
Egyptian females (1987)	4960	1606 (72)	838 (43)	499 (25)	62.6 (4.4)
Indian males (farmers) (1980)	13	1576 (17)	—	—	44.6 (1.4)
Central Indian male farm workers (1989)	39	1620 (50)	739 (26)	509 (30)	49.3 (6.0)
South Indian males (workers) (1992)	128	1607 (60)	791 (40)	542 (38)	56.6 (5.1)
Indonesian females (1985)	468	1516 (54)	719 (34)	—	—
Indonesian males (1985)	949	1613 (56)	872 (37)	—	—
Irish males (1991)	164	1731 (58)	911 (30)	508 (28)	73.9 (8.7)
Italian females (1991)	753	1610 (64)	850 (34)	495 (30)	58 (8.3)
Italian males (1991)	913	1733 (71)	896 (36)	541 (30)	75 (9.6)
Jamaican females (1991)	123	1648	832	—	61.4
Jamaican males (1991)	30	1749	856	—	67.6
Korean female workers (1989)	101	1580 (57)	833 (32)	460 (22)	53.9 (6.9)
Malay females (1988)	32	1559 (66)	831 (39)	—	—
Saudi-Arabian males (1986)	1440	1675 (61)	—	—	—
Singapore males (pilot trainees) (1995)	832	1685 (53)	894 (32)	—	—
Sri Lankan females (1985)	287	1523 (59)	774 (22)	—	—
Sri Lankan males (1985)	435	1639 (63)	833 (27)	—	—
Sudanese males					
Villagers (1981)	37	1687 (63)	—	—	57.1 (7.6)
City dwellers (1981)	16	1704 (72)	—	—	62.3 (13.1)
City dwellers (1982)	48	1668	—	—	51.3
Soldiers (1981)	21	1735 (71)	—	—	71.1 (8.4)
Soldiers (1982)	104	1728	—	—	60.0
Thai females (1991)	250	1512 (48)	—	—	—
	711	1540 (50)	817 (27)	—	—

Table 1-11.   (continued)

	Sample size N	Stature	Sitting height	Knee height, sitting	Weight
Thai males (1991)	250	1607 (20)	—	—	—
	1478	1654 (59)	872 (32)	—	—
Turkish females (1991)					
Villagers	47	1567 (52)	792 (38)	486 (27)	69.1 (13.8)
City dwellers	53	1563 (55)	786	471	65.9 (13.0)
Turkish males (soldiers) (1991)	5108	1702 (60)	888 (34)	513 (28)	63.3 (7.3)
Vietnamese (American V.) (1993)					
Females	30	1559 (61)	—	—	48.6
Males	41	1646 (54)	—	—	58.9
U.S. Midwest workers, with shoes and light clothes (1993)					
Females	125	1637 (62)	—	—	64.7 (11.8)
Males	384	1778 (73)	—	—	84.2 (15.5)
U.S. male miners (1993)	105	1803 (65)	—	—	89.4 (15.1)

Statistical Body Models

It is often desired to combine body dimensions to construct a complete model of the human body or of its major components. For this purpose, two methods have been employed: one is incorrect, but the other is suitable.

The percentile statistic is convenient for establishing the location of one given datum measured along its continuum range, for example, 1528 mm as the 5th percentile value for female stature. However, it is false to believe that all other body component measures of that person must also be at the 5th percentile; $p5$ leg length plus $p5$ torso length plus $p5$ head height do not add up to $p5$ stature (Robinette and McConville, 1981): a person of $p5$ stature may have relatively short legs but a long torso, or vice versa.

Regression equations, however, are suitable to generate discrete body measures, using as predictor variables other dimensions whose values are known for the population sample of interest. Given that the equations are proper, predicted values can be added or subtracted. For example, 5th percentile values for leg, trunk, and head heights predicted from regressions do add up to the correct $p5$ stature.

Hence, percentiles are useful descriptors of discrete body measures, but they can be "stacked" only if derived from regression equations concerning the sample in question.

How to Get Missing Data

In many cases, we do not know the exact body sizes of persons for whom we want to design products, tools, and equipment. This is not a great problem if the

Table 1-12. Average anthropometric data in millimeters estimated for 20 regions of the earth. Adapted from Juergens, Aune, and Pieper (1990).

	Stature		*Sitting height*		*Knee height, sitting*	
	Females	**Males**	**Females**	**Males**	**Females**	**Males**
North America	1650	1790	880	930	500	550
Latin America						
Indian Population	1480	1620	800	850	445	495
European and						
Negroid population	1620	1750	860	930	480	540
Europe						
North	1690	1810	900	950	500	550
Central	1660	1770	880	940	500	550
East	1630	1750	870	910	510	550
Southeast	1620	1730	860	900	460	535
France	1630	1770	860	930	490	540
Iberia	1600	1710	850	890	480	520
Africa						
North	1610	1690	840	870	500	535
West	1530	1670	790	820	480	530
Southeast	1570	1680	820	860	495	540
Near East	1610	1710	850	890	490	520
India						
North	1540	1670	820	870	490	530
South	1500	1620	800	820	470	510
Asia						
North	1590	1690	850	900	475	515
Southeast	1530	1630	800	840	460	495
South China	1520	1660	790	840	460	505
Japan	1590	1720	860	920	395	515
Australia						
European extraction	1670	1770	880	930	525	570

product is similar to items already in use and if the users are fairly well known to us, such as our colleagues. In this case, we can probably take a few measurements of our acquaintances to "guestimate" what the needed dimensions might be.

Often, however, more than such informal information gathering is necessary. Two avenues are open: One is to apply statistical procedures to deduce from existing data those that we need to know; the other is to conduct a formal

Table 1-13.   Values of q for sample size determination

q	**Statistic of Interest**
1.96	Mean
1.39	Standard deviation
2.46	50th percentile
2.46	45th and 55th percentile
2.49	40th and 60th percentile
2.52	35th and 65th percentile
2.58	30th and 70th percentile
2.67	25th and 75th percentile
2.80	20th and 80th percentile
3.00	15th and 85th percentile
3.35	10th and 90th percentile
4.14	5th and 95th percentile
4.46	4th and 96th percentile
4.92	3rd and 97th percentile
5.67	2nd and 98th percentile
7.33	1st and 99th percentile

anthropometric survey. Doing a survey is a major enterprise and is best left to qualified anthropometrists, but a few general remarks may help the planning (Kroemer, 1989; Roebuck, 1995).

First, we have to identify the people of interest. Next, we must decide on how to select a representative sample. That decision includes the determination of the sample size. For a variety of reasons, we usually wish to keep the sample as small as possible. Assuming normal distribution of the measured variable, the required sample size N can be estimated from

$$N = q^2 * S^2 * d^{-2}$$

where S is the estimated standard deviation of the data, d is the desired accuracy (in $\pm d$ units) of the measurement, and q is taken from Table 1-13. (Note that the multiplier values q depend on the statistic of interest.)

If the initially calculated sample size N is below 100, the values for q given in Table 1-12 should be replaced by

$$q = 2.00 \text{ for } 100 > N > 4$$

$$q = 2.05 \text{ for } 40 > N > 20$$

$$q = 2.16 \text{ for } 20 > N > 10$$

$$q = 2.78 \text{ for } 10 > N$$

and the calculation should be repeated (Roebuck, Kroemer, and Thomson, 1975).

There are several statistical procedures for estimating data from existing related information.

Estimation by Ratio Scaling

Ratio scaling (as used by Pheasant, 1986, 1996) is one technique for estimating data from known dimensions. It relies on the assumption that, though people vary greatly in size, they are likely to be relatively similar in proportions. This holds true for body components that are interrelated to each other, as discussed in detail by Roebuck, Kroemer, and Thomson (1975), by Pheasant (1982) and, most recently, by Roebuck (1995). For example, many body lengths are highly intercorrelated with each other; also, many body breadths are related, as are circumferences. However, it is not true that all body lengths (or breadths or circumferences) are highly correlated, and certainly many lengths are not closely related to breadths, and breadths are not closely related to many circumferences. Thus, we have to be very careful in deriving one set of data from another.

It is recommended to use, for any ratio scaling, only pairs of data that are related to each other with a coefficient of correlation of at least 0.7. This assures that the variability of the derived information is at least 50% determined by the variability of the predictor. (This derives from the square of the correlation coefficient: $0.7^2 = 0.49$.) Never use ratio scaling if you must assume that the sample you use to scale has body proportions different from those of the other set; for example, many Asian populations have proportionally shorter legs and longer trunks than Europeans or North Americans.

For sets of highly correlated data, we can establish an estimate E for the ratio scaling factor of a desired dimension in the population sample Y (d_y): If we know the value of that dimension in sample X (d_x) and if we know the values of the source reference dimension in both samples X and Y (D_x and D_y):

$$\frac{d_x}{D_x} = E = \frac{d_y}{D_y} \tag{1-2}$$

With $E = d_x/D_x$ known, we can calculate the desired dimension

$$d_y = E * D_y \tag{1-2a}$$

in stepwise fashion, as shown in the following.

Step 1: In population sample X, establish the scaling factor E between the desired dimension and a known reference dimension. The reference parameter must be common for both population samples.

Example:

On average, East German men have an eye height (d_x) of 160.1 cm, and their average stature (D_x) is 171.5 cm. If West Germans have a mean stature (D_y) of about 175 cm, what is their approximate mean eye height?

$$E = \frac{d_x}{D_x} = \frac{160.1}{171.5} = 0.933.$$

Step 2: With E now known, the desired unknown dimension in population sample Y equals E times the reference parameter in sample Y.

Example:

Eye height in sample Y equals $E *$ stature in sample Y

$$dy = E * D_y = 0.933 * 175 \text{ cm} = 163.3 \text{ cm} \qquad (1\text{-}3)$$

For practical reasons, the common parameter is often stature. Note, however, that stature generally relates well with other heights but not necessarily with depths, breadths, circumferences, or weight (as discussed earlier). Thus, ratio scaling must be done with great caution and careful consideration of the circumstances, taking into account especially statistical correlations. Ratio scaling equations have not been justified mathematically but have been used primarily for expediency.

For more detail on ratio scaling, see Pheasant (1986, 1996) and Roebuck (1995). The technique of ratio scaling has been applied primarily to estimate the mean of a required dimension and its standard deviation.

Estimation by Regression Equations

Another way to estimate the relations among dimensions is through regression equations. Most regression equations are bivariate, meaning that two variables are involved, and it is presumed that the two variables are linearly related to each other. The general form is

$$y = a + bx \qquad (1\text{-}4)$$

where x is the known mean value and y the predicted mean. The constants a (the intercept) and b (the slope) must be determined (known) for the data set of interest. A recent example of this procedure is the estimation of body dimensions of American soldiers by Cheverud et al. (1990).

If you predict the mean value of y (for any value of x) using the regression equation (1-4), you must expect that the actual values of y are scattered about the mean in a normal (Gaussian) probability distribution. The standard error SE of the estimate depends on the correlation r between x and y (and the standard deviation of y, S_y) according to

$$SE_y = S_y (1 - r^2)^{1/2} \qquad (1\text{-}5)$$

Note that the assumption of linearity is usually made but often neither clearly stated nor verified. Roebuck (1995) discussed the implications in some detail, including the extension of this concept to develop multivariate regression equations, as well as principal component analyses and boundary description analyses.

Estimation by Probability Statistics

In most cases, we are unable to measure every person with respect to size or strength. If we were able to do so, we would describe the parameters of that total population by the mean (average) and standard deviation, designated by the Greek letters μ and σ. (This is the terminology convention used in most statistics books.) In reality, we can measure only a subgroup (sample) and, from its parameters, we infer or estimate what the actual population would have yielded. We say that

$$m = \frac{(\Sigma x)}{n} \qquad (1\text{-}6)$$

where m is the mean (average), x the individual measurement, and n the number of measured individuals. The distribution of the data is described by the equation

$$s = \frac{\sqrt{\Sigma(x - m)^2}}{n} \qquad (1\text{-}6a)$$

with s the standard deviation of the sample. If the sample size n is small (by convention, 30 or less), we make the arbitrary correction

$$s = \frac{\sqrt{\Sigma(x - m)^2}}{n - 1} \qquad (1\text{-}6b)$$

The smaller n is, the larger the standard error SE in sampling:

$$SE \text{ of the mean } m = \frac{s}{\sqrt{n}} \qquad (1\text{-}7)$$

$$SE \text{ of the standard deviation } s = \frac{s}{\sqrt{2n}} = 0.71 \, SE_m \qquad (1\text{-}8)$$

As the number n increases, the mean m and the standard deviation s become more reliable estimates of the underlying general population, that is, of μ and σ.

It is often useful to describe the variability of a sample by dividing the standard deviation s by the mean m (and multiplying the result by 100). This yields the coefficient of variation CV (in percent),

$$CV = 100 \frac{s}{m} \qquad (1\text{-}9)$$

This expression is independent of the magnitude and of the unit of measurement.

Groups of human measurements show characteristic variabilities. Typical coefficients of variation are listed in Table 1-14. This information can be used to judge the reliability of reported data.

Table 1-14.   Variability of body measurements.

Variables measured	CV, %
Body heights (stature, sitting height, elbow height, etc.)	3–5
Body breadths (hip, shoulder, etc.)	5–9
Body depths (abdomen, chest, etc.)	6–9
Reaches	4–10
Total body weights	10–20
Joint ranges	7–30
Muscular static strength	10–85

Combining Anthropometric Data Sets

Occasionally, we must add or subtract anthropometric values; for example, total arm length is the sum of upper and lower arm lengths.

If you want to add two measures, such as leg length and torso (with head) length, you generate a new combined distribution, stature. In doing so, you must take into account the covariation (COV) between the two measures of leg and torso: usually (but not always), a taller torso is associated with a taller head. This is mathematically described by the correlation coefficient r between the two data sets x and y and their standard deviations s_x and s_y:

$$COV(x,y) = r_{x,y} * s_x * s_y \qquad (1\text{-}10)$$

This allows calculation of the *sum* of the two mean values of the x and y distributions from

$$m_z = m_x + m_y \qquad (1\text{-}11)$$

and the estimated standard deviation of z from

$$S_z = [s_x^2 + s_y^2 + 2rs_x s_y]^{1/2} \qquad (1\text{-}12)$$

The *difference* between two mean values is

$$m_z = m_x - m_y \qquad (1\text{-}13)$$

and its standard deviation

$$S_z = [s_x^2 + s_y^2 - 2rs_x s_y]^{1/2} \qquad (1\text{-}14)$$

Example A:

What is the 95p shoulder-to-fingertip length? The mean lower arm *LA* link length (with the hand) is 442.9 mm with a standard deviation of 23.4 mm. The mean upper arm *UA* link length is 335.8 mm and its standard deviation is 17.4 mm.

The multiplication factor of $k = 1.65$ (from Table 1-8) leads you to the 95th percentile. But you cannot calculate the sum of the two 95p lengths because you would be disregarding their covariance; instead, you calculate the sum of the mean values first [see eq. (1-11)].

$$m = m_{LA} + m_{UA} = 442.9 + 335.8 = 778.7 \text{ mm}$$

The standard deviation is calculated next, using an assumed coefficient of correlation of 0.4 [see eq. (1-12)]

$$s = [23.4^2 + 17.4^2 + 2 * 0.4 * 23.4 * 17.4]^{1/2} \text{ mm}$$

$$s = 34.3 \text{ mm}$$

The 95p total arm length AL can now be calculated [see eq. (1-1) on page 28]

$$AL_{95} = 778.7 \text{ mm} + 1.65 * 34.3 \text{ mm} = 835.3 \text{ mm}$$

Example B:

What is the average arm (acromion-to-wrist) length of an American pilot? For a standing pilot, the 90th percentile acromial (shoulder) height is 1532.0 mm and the wrist height is 905.6 mm; for the 10th percentile, the values are 1379.5 and 808.6 mm, respectively. The correlation between shoulder and wrist heights is estimated at 0.3.

You first calculate the mean acromion (*A*) and wrist (*W*) heights to be able to estimate the standard deviations.

$$m_A = \frac{(1532.0 + 1379.5) \text{ mm}}{2} = 1455.75 \text{ mm}$$

With $k = 1.28$ taken from Table 1-8 [see eq. 1-1]

$$s_A = \frac{(1532.0 - 1455.75) \text{ mm}}{1.28} = 59.6 \text{ mm}$$

or

$$s_A = \frac{(1455.75 - 1379.5) \text{ mm}}{1.28} = 59.6 \text{ mm}$$

Likewise, [see eqs. (1-11 and 1-12)]

$$m_W = \frac{(905.6 + 808.6) \text{ mm}}{2} = 857.1 \text{ mm}$$

$$s_W = \frac{(905.6 - 857.1) \text{ mm}}{1.28} = 37.9 \text{ mm}$$

or $\qquad s_W = \dfrac{(857.1 - 808.6) \text{ mm}}{1.28} = 37.9 \text{ mm}$

The average arm length (acromion-to-wrist, AW) is [see eq. (1-13)]

$$m_{AW} = m_A - m_W = 1455.75 \text{ mm} - 857.1 \text{ mm} = 598.65 \text{ mm}$$

The standard deviation of the arm length is [see eq. (1-14)]

$$s_{AW} = [59.6^2 + 37.9^2 - 2 * 0.3 * 59.6 * 37.9]^{1/2} \text{ mm} = 60.3 \text{ mm}$$

Example C

What is the mass of the torso of a $75p$ female? The estimated mass of the torso and head combined has a mean of 35.8 kg and a standard deviation of 5.2 kg. The estimated mass of the head, measured separately, has a mean of 5.8 kg with a standard deviation of 1.2 kg. Assume the correlation between head and torso to be 0.1.

The mean torso mass is the difference [see eq. (1-13)]

$$\text{mean}_{\text{torso}} = 35.8 \text{ kg} - 5.8 \text{ kg} = 30.0 \text{ kg}$$

The standard deviation is calculated from [see eq. (1-14)]

$$s_{\text{torso}} = [5.2^2 + 1.2^2 - 2 * 0.1 * 5.2 * 1.2]^{1/2} - \text{kg} = 5.2 \text{ kg}$$

The mass of a 75th percentile torso is (with $k = 0.67$ taken from Table 1-8) [see eq. (1-1)]

$$\text{mass}_{\text{torso}} \, 75p = 30.0 \text{ kg} + 0.67 * 5.2 \text{ kg} = 33.5 \text{ kg}$$

Estimating the Standard Deviation

Often, one has only mean values of body dimensions but no information about the associated standard deviations. Pheasant (1986, 1996) developed the following equations for Americans, based on data provided by Roebuck, Kroemer, and Thompson (1975), to estimate the standard deviation s from the mean value m for most heights, lengths, and breadths (all data in millimeters):

$$\text{Men: } s \approx (57.030 * 10^{-3}) \, m - (8.347 * 10^{-6}) \, m^2 \qquad (1\text{-}15)$$

$$\text{Women: } s \approx (57.830 * 10^{-3}) \, m - (10.347 * 10^{-6}) \, m^2 \qquad (1\text{-}16)$$

For sitting elbow height, thigh thickness, hip breadth, and depth:

$$\text{Men: } s \approx 7.864 + (69.770 * 10^{-3}) \, m \qquad (1\text{-}17)$$

$$\text{Women: } s \approx 4.249 + (94.670 * 10^{-3}) \, m \qquad (1\text{-}18)$$

For a more detailed discussion of this and other estimating procedures, consult the 1995 book by Roebuck.

Composite Population

It may be necessary to consider a population that consists of two distinct and known subsamples. An example is to design for a user group that consists of $a\%$ females and $b\%$ males, with $a + b = 100\%$. To determine at what percentile of the composite population a specific value of x is, we proceed stepwise as follows (Kroemer, 1983):

Step 1: Determine k factors associated with x in the samples a and b. For sample a,

using $\qquad\qquad\qquad p_a = m_a + k_a S_a \qquad\qquad\qquad (1\text{-}1)$

yields $\qquad\qquad\qquad k_a = (p_a - m_a) \, S_a^{-1} \qquad\qquad\qquad (1\text{-}19)$

Similarly, for sample b,

$$k_b = (p_b - m_b) \, S_b^{-1} \qquad (1\text{-}20)$$

Step 2: Obtain factor k associated with x in the combined population:

$$k = a \, k_a + b \, k_b \qquad (1\text{-}21)$$

Step 3: Determine percentile p associated with k, using Table 1-8.
Step 4: If percentiles p for each x are known in both groups, we may simply add the proportioned percentiles:

$$p = a \, p_a + b \, p_b \qquad (1\text{-}22)$$

The "Normative" Adult

We have become used to design for the "regular" adult, who is presumed to be of "normal" anthropometry, that is, to have body dimensions, such as stature, hand reach, or weight, all close to mean values; to possess "normal" physiological functions, such as of the metabolic, circulatory, and respiratory subsystems; and to have "average" nervous control, sensory capabilities, and intelligence. In fact, there are ergonomic texts that use as their model the "average male" of North America.

Thus, by default or for reasons, the normative stereotype of many human factor engineers is the regular adult (woman or man) who is physically and psychologically healthy and able and willing to perform. This mythical being has become the reference population prototype to which we compare other subgroups, such as children, temporarily or permanently impaired persons, women during pregnancy, or aging people. Kroemer, Kroemer, and Kroemer-Elbert (1994) discuss how to design for these special populations that deviate in size, strength, or other performance capabilities from the normative adult.

Motions and Postures

The human is unable to maintain a given posture over long periods of time. Standing still, sitting still, and even lying still quickly become uncomfortable and then physically impossible with time; if enforced by injury or sickness, circulatory and metabolic functions become impaired, and bed sores appear. The human body is made to move.

Unfortunately, the erect (or upright) posture has been employed as a design model, probably because it is easily visualized and made into a design template. This upright idol was promoted by orthopedists of the late 19th century, who translated their postural concerns into the desire for an erect trunk posture, especially when sitting in school or office. Yet, designing to fit motion ranges instead of fixed postures is not difficult.

Mobility

The human body has various degrees of freedom to move in its articulations. Maximal angular displacements in major body joints are shown in Chapter 2. "Convenient" mobility is somewhere within the range of maximal values—in some cases, in the midrange and, in others, near extremes. Habits and skill, as well as strength requirements, may make different ranges preferred.

Our bodies are designed for movement especially in the arms, with shoulder and elbow joints providing extensive angular freedom. The legs are able to propel the body on the ground, with major motions occurring in the knee and hip joints. Movements of the trunk occur mostly in flexion and extension at the lower back. However, these bending and unbending motions (in the medial plane) are rather limited and often lead to overexertions, especially if combined with sideways twisting of the torso; low back pain has been reported throughout the history of mankind. Wrist problems have been associated with excessive motion requirements since the early 1700s. Our head and neck have limited mobility in bending and twisting. Our thumbs and fingers, as well as eyes, have limited but finely controlled motion capability.

Actual ranges of motion (also called mobility or flexibility) depend much on age, health, fitness, training, and skill. Mobility ranges have been measured on dissimilar groups of people with various measuring instructions and techniques; hence, there is much diversity in reported results.

Design for Movements

Design for motions starts by establishing the "extremes" of the expected movements. "Convenient" movement ranges may cluster around the mean of mobility in a body joint or may be close to a flexibility limit. For example, a person walking about on a job, or standing, has the knees nearly extended most of the time, that is, with the knees angled close to the extreme value of about 180 deg. The sagittal hip angle (between trunk and thigh) is also in the neighborhood of 180 deg. Both angles change to about 90 deg when sitting. While sitting or moving about, the trunk is held normally fairly close to erect, as are neck and head. In most work situations, the upper arm hangs from the shoulder while the elbow angle tends to be near 90 deg; but the wrist is best held straight.

Designing for motion is most easily achieved by selecting the appropriate ranges of angulation in the major body joints involved. Then, body dimensions reported for the standardized postures are adjusted to accommodate the motion range. For convenience, this range can be depicted as the area between two positions, such as knee angles ranging between 60 and 105 deg, or as a motion envelope, circumscribed by combined hand-and-arm movement, or by the clearance envelope under (within, through, beyond) which body parts must fit. Table 1-15 indicates mobility ranges at work.

Criteria for work space layout include those related to human strength, speed, accuracy, effort, and the traditional "rational design principles" related to importance, function, frequency, and sequence, as listed in Table 1-16. Achieving the task while assuring safety for the human and avoiding overuse and unnecessary effort are primary goals.

Basic work space design faults should be avoided. These include:

1. *Avoid twisted body positions,* especially of the trunk and neck. This results often from poor location of work objects, controls, and displays.
2. *Avoid forward bending of trunk, neck, and head.* This is frequently provoked by improperly positioned controls and visual targets, including working surfaces that are too low.
3. *Avoid postures that must be maintained* for long periods of time, especially at the extreme limits of the range of motion. This is particularly important for the wrist and the back.
4. *Avoid holding the arms raised.* This commonly results from locating controls or objects too high, that is, higher than the elbow when the upper arm hangs down. The upper limit for regular manipulation tasks is about chest height.

Table 1-15. Actual motion ranges of the body.

		Estimated predominant motion ranges	
Major joint mobilities		**Walking (standing)**	**Sitting**
Ankle	Ext/flexion	About midrange (70–100 deg) (M)	
Knee	Ext/flexion	Nearly fully extended (160–180 deg) (E)	About right angle (70–100 deg)
Hip	Ext/flexion	Nearly fully extended (170–190 deg) (E)	About right angle (80–100 deg)
Spine	Ext/flexion	Nearly fully extended but occasionally flexed from full extension by about 45 deg (E)	
	Lateral twist	Normally no twist, but occasionally twisted left or right by about 20 deg (E)	
Head and neck	Ext/flexion	Nearly fully extended but occasionally flexed from full extension by about 20 deg (E)	
	Lateral twist	Normally no twist, but occasionally twisted left or right by about 20 deg (E)	
Shoulder	Ext/flexion	Usually neither flexed nor extended (upper arm hangs down) but occasionally flexed by about 30 deg	
	Ab-/abduction	Usually neither adducted nor abducted but occasionally moved ± 30 deg	
Elbow	Ext/flexion	Often flexed at about 90 deg but with frequent motions through often large ranges (170–40 deg)	
Forearm	Pro/supination (twist)	Normally neither pronated nor supinated but occasionally twisted ± 20 deg (M)	
Wrist	Ext/flexion Ad-/abduction (radial/ulnar deviation)	Normally about straight but often extended/flexed by about 20 deg and laterally bent by about 10 deg (M)	

Note: Only ankle flexion/extension and forearm and wrist motions are commonly at midrange (M), while many other motions occur in other ranges, often near extremes (E).

Reach Envelopes

Preferred work areas of the hands and feet are in front of the body, within curved envelopes that reflect the mobility of the forearm in the elbow joint or of the total arm in the shoulder joint, of the lower leg in the knee joint and of the total leg in the hip joint. Thus, these envelopes are often described as (partial) spheres around the presumed locations of the body joints. However, the preferred ranges within the possible motion zones are different when the main requirements are

Table 1-16. Guidelines for Workspace Design

Human strength	facilitate extension of strength (work, power) by object location and orientation.
Human speed	place items so that they can be reached and manipulated quickly.
Human effort	arrange work so that it can be performed with least effort.
Human accuracy	select and position objects so that they can be manipulated and seen with ease.
Importance	the most important items should be in the most accessible locations.
Frequency of use	the most frequently used items should be in the most accessible locations.
Function	items with similar functions should be grouped together.
Sequence of use	items which are commonly used in sequence should be laid out in that sequence.

strength, speed, accuracy, or vision (as discussed, in some detail, by Kroemer, Kroemer, and Kroemer-Elbert, 1994). Thus, there is not one reach envelope, but different preferred envelopes exist.

For each job situation, the ergonomic designer determines the dominant requirements of the task, for example whether the operator works while sitting or walking (standing), performs wide-ranging or specialized work, must exert large or small forces, can execute fast and gross or slow and exact motions, or needs high or low visual control. Such circumstances affect the selection of the specific work envelope.

SUMMARY

Traditional techniques for measuring the human body use a set of simple tools based mostly on grids, rods, and tapes as measuring scales. New developments are using photographic or laser-based three-dimensional methods. Measurements are taken preferably to solid identifiable landmarks of the skeleton. Other measurements are statistically related to these basic dimensions.

Statistical changes in body dimensions are due to secular body size development, aging, changes in health and fitness, local population composition, and immigration. The influence of such changes on the population anthropometry in general is only subtle. In fitting equipment to a defined population subgroup, however, the engineer must consider the body dimensions that are specific to the group.

Most currently available data were measured on (male) soldiers. Data describing civilians are scarce and are often derived from military data.

Body proportions differ greatly from person to person. Hence, body types or the construct of the "average person" (or of other single-percentile phantoms) are not suitable means to assess body dimensions or capabilities for engineering purposes.

It is inexcusable to design tasks, tools, or workstations for the phantom "average people." No such persons exist, and design for the average fits nobody. Instead, ranges of body sizes, of motions, and of strengths establish the design criteria. This is easy to do for the designer and engineer who starts with proper anthropometric information and applies it ergonomically, that is, with ease and efficiency as the guiding principles.

REFERENCES

Andres, R. (1984). Mortality and Obesity: The Rationale for Age-Specific Height-Weight Tables. In Andres, R., Bierman, E.L., and Hazzard, W.R. (eds.). *Principles of Geriatric Medicine.* New York, NY: McGraw-Hill, pp. 311–318.

Chaffin, D.B., and Andersson, G.B.J. (1984). *Occupational Biomechanics,* 3rd ed. New York, NY: Wiley.

Chapanis, A. (ed.) (1975). *Ethnic Variables in Human Factors Engineering.* Baltimore, MD: John Hopkins University Press.

Cheverud, J., Gordon, C.C., Walker, R.A., Jacquish, C., Kohn, L., Moore, A., and Yamashita, N. (1990). *1988 Anthropometric Survey of U.S. Army Personnel: Correlation Coefficients and Regression Equations.* Natick TR 90/032-6. Natick, MA: U.S. Army Research, Development and Engineering Center.

Garrett, J.W., and Kennedy, K.W. (1971). *A Collation of Anthropometry.* AMRL-TR-68-1, Wright-Patterson AFB, OH: Aerospace Medical Research Laboratories.

Gordon, C.C., Churchill, T., Clauser, C.E., Bradtmiller, B., McConville, J.T., Tebbetts, I., and Walker, R.A. (1989). *1988 Anthropometric Survey of U.S. Army Personnel: Summary Statistics Interim Report.* Natick TR-89/027. Natick, MA: U.S. Army Natick Research, Development and Engineering Center.

Fluegel, F., Greil, H., and Sommer, K. (1986) *Anthropologischer Atlas.* Berlin, Germany: Tribuene.

Hertzberg, H.T.E. (1968). The Conference on Standardization of Anthropometric Techniques and Terminology. *American Journal of Physical Anthropology* 28:1–16.

Juergens, H.W., Aune, I.A., and Pieper, U. (1990). *International Data on Anthropometry.* Occupational Safety and Health Series No. 65. Geneva, Switzerland: International Labour Office.

Kagimoto, Y., (ed.) (1990). Anthropometry of JASDF Personnel and Its Applications for Human Engineering. Tokyo, Japan: Aeromedical Laboratory, Air Development and Test Wing JASDF.

Kennedy, K.W. (1986). *A Collation of United States Air Force Anthropometry.* AMRL TR 85-062. Wright-Patterson AFB, OH: Aerospace Medical Research Laboratory.

Kroemer, K.H.E. (1983). Engineering Anthropometry. In Oborne, D.J., and Gruneberg, M.M. (eds.) *The Physical Environment at Work.* London, UK: Wiley, pp. 39–68.

Kroemer, K.H.E. (1989). Engineering Anthropometry. *Ergonomics,* 32:767–784.

Kroemer, K.H.E., Kroemer, H.B., and Kroemer-Elbert, K.E. (1994). *Ergonomics: How to Design for Ease and Efficiency.* Englewood Cliffs, NJ: Prentice-Hall.

Kroemer, K.H.E., Kroemer, H.J., and Kroemer-Elbert, K.E. (1990). *Engineering Physiology: Bases of Human Factors/Ergonomic,* 2nd ed. New York, NY: Van Nostrand Reinhold.

Lohman, T.G., Roche, A.F., and Martorell, R. (eds.) (1988). *Anthropometric Standardization Reference Manual.* Champaign, IL: Human Kinetics.

McConville, J.T., Robinette, K.M., and Churchill, T. (1981). *An Anthropometric Data Base for Commercial Design Applications.* Final Report, NSF DAR-80 09 861. Yellow Springs, OH: Anthropology Research Project, Inc.

NASA (ed.) (1978). *Anthropometric Sourcebook* (3 vols.) NASA Reference Publication 1024. Houston, TX: L.B.J. Space Center, NASA: (NTIS, Springfield, VA 22161, Order 79 11 734).

Pheasant, S.T. (1982). A Technique for Estimating Anthropometric Data from the Parameters of the Distribution of Stature. *Ergonomics* 25:1 981–992.

Pheasant, S. (1986). *Bodyspace,* 1st ed. London, UK: Taylor & Francis.

Pheasant, S. (1996). *Bodyspace: Anthropometry, Ergonomics and the Design of Work,* 2nd ed. London, UK: Taylor and Francis.

Robinette, K.M., and McConville, J.T. (1981). *An Alternative to Percentile Models.* SAE Technical Paper 810217. Warrendale, PA: Society of Automotive Engineers.

Roebuck, J.A. (1995). *Anthropometric Methods: Designing to Fit the Human Body.* Santa Monica, CA: Human Factors and Ergonomics Society.

Roebuck, J.A., Kroemer, K.H.E., and Thomson, W.G. (1975). *Engineering Anthropometry Methods.* New York, NY: Wiley.

White, R.M. (1982). *Comparative Anthropometry of the Foot.* Natick TR 83-010. Natick, MA: United States Army Natick Research and Development Laboratories.

FURTHER READING

Asimov, I. (1963). *The Human Body: Its Structure and Operation.* New York, NY: New American Library/Signet.

Chapanis, A. (ed.) (1975). *Ethnic Variables in Human Factors Engineering.* Baltimore, MD: John Hopkins University Press.

Gould, S.J. (1981). *The Mismeasure of Man.* New York, NY: Norton.

NASA (ed.) (1978). *Anthropometric Sourcebook* (3 vols.) NASA Reference Publication 1024. Houston, TX: NASA (NTIS, Springfield, VA 22161, Order No. 79 11 734).

Kroemer, K.H.E., Kroemer, H.B., and Kroemer-Elbert, K.E. (1994). *Ergonomics: How to Design for Ease and Efficiency.* Englewood Cliffs, NJ: Prentice-Hall.

Roebuck, J.A. (1995). *Anthropometric Methods. Designing to Fit the Human Body.* Santa Monica, CA: Human Factors and Ergonomics Society.

Appendix A

Correlation Table

Table 1–17 is compiled from data on male and female U.S. soldiers measured in the 1970s.* It shows the coefficients of correlation between 19 body dimensions important for design purposes.

> Standing heights: acromion, cervical, crotch
> Sitting heights: sitting, eye, knee, popliteal
> Sitting depths, breadths: buttock-knee, popliteal, hip
> Forward reach: thumb-tip
> Circumferences: chest, waist
> Hand dimensions: breadth, circumference, length
> Head dimensions: breadth, circumference, length

Nine body dimensions are selected for their predictive powers, that is, their high correlations with certain important design variables. Listed for each variable are the correlations for three male samples (top three rows), followed by the data for two female samples.

The table indicates that there are generally high correlations within heights, within certain circumferences, within measurements on the hand, and within some head dimensions.

Stature is a valuable predictor for many selected design variables, that is, for (standing) acromion, cervicale, and crotch height and, for (sitting) height, and eye, knee, popliteal height; even buttock-popliteal length, thumb-tip reach, and hand length are rather well correlated. Waist height is a good predictor for trunk heights and for leg lengths, while weight is related primarily to trunk measurements contained in the table. These statements hold true, with only minor shifts, for male and female soldiers.

---

*For more information, see the 1990 edition of this book.

Table 1-17. Correlations between (19) design dimensions and (9) predictor dimensions.

Design Dimensions	Stature	Bitragion Breadth	Head Length	Heel-Ankle Circumference	Hip Breadth, standing	Palm Length	Shoulder Circumference	Waist Height	Weight
Acromion	.960**	.218	.260	.599	.397	.575	.306	.911**	.479
Height,	.958**	.207	.224	.622*	.448	.525	.344	.906**	.554
standing	.959**	.184	.228	.641*	.425	.527	.304	.909**	.480
	.959**	.237	.298	na	.350	na	.350	.919**	.552
	.968**	na	.336	.618*	na	.574	.354	.891**	.567
Cervical Height,	.977**	.210	.263	.611*	.402	.586	.324	.929**	.482
standing	.977**	.185	.217	.630*	.427	.526	.331	.931**	.529
	.978**	.172	.237	.655*	.420	.545	.308	.919**	.476
	.977**	.221	.299	na	.349	na	.334	.927**	.540
	.979**	na	.320	.599	na	na	.375	.894**	.563
Crotch Height,	.839**	.075	.191	.482	.196	.538	.141	.887**	.252
standing	.856**	.082	.170	.517	.283	.484	.212	.905**	.359
	.857**	.082	.196	.601*	.229	.549	.175	.929**	.312
	.849**	.174	.280	na	.225	na	.264	.909**	.430
	.861**	na	.320	.599	na	.629*	.247	.882**	.402
Sitting Height	.778**	.195	.270	.487	.403	.388	.296	.537	.434
	.786**	.158	.244	.495	.376	.367	.291	.580	.467
	.732**	.196	.227	.404	.422	.275	.289	.476	.412
	.801**	.204	.275	na	.371	na	.312	.607*	.091
	.767**	na	.261	.350	na	.304	.253	.570	.421
Eye Height,	.753**	.173	.233	.461	.388	.374	.281	.518	.411
sitting	.738**	.121	.193	.456	.346	.341	.260	.544	.412
	.709**	.170	.184	.383	.400	.273	.265	.463	.380
	.737**	.170	.229	na	.371	na	.287	.562	.448
	.738**	na	.220	.331	na	.279	.235	.551	.399
Knee Height,	.878**	.216	.258	.618*	.344	.612*	.331	.903**	.460
sitting	.882**	.193	.221	.665*	.428	.539	.358	.904**	.539
	.873**	.177	.253	.738**	.420	.583	.373	.892**	.524
	na	na	na	na	na	na	na	na	na
	.857**	na	.362	.694*	na	.647*	.363	.850**	.546
Popliteal Height,	.808**	.036	.170	.427	.030	.540	.020	.832**	.090
sitting	.841**	.106	.175	.529	.221	.513	.182	.883**	.299
	.830**	.079	.206	.615*	.203	.569	.176	.888**	.289
	.728**	.149	.253	na	.181	na	.230	.762**	.370
	.847**	na	.326	.622*	na	.651*	.252	.853**	.401
Buttock-Knee	.801**	.201	.244	.567	.480	.504	.434	.844**	.582
Length, sitting	.760**	.229	.208	.586	.550	.454	.435	.790**	.636*
	.766**	.224	.249	.682*	.554	.487	.493	.795**	.662*
	.769**	.245	.296	na	.527	na	.482	.809**	.694*
	.761**	na	.342	.671*	na	.539	.506	.748**	.720**
Buttock-Popliteal	.684*	.145	.303	.131	.341	.415	.284	.737**	.397
Length, sitting	.686*	.161	.157	.499	.506	.374	.394	.718**	.565
	.706**	.160	.226	.615*	.491	.446	.430	.748**	.586
	.653*	.197	.249	na	.426	na	.391	.710**	.565
	na	na	na	na	na	na	na	na	na
Thumb-Tip	.627*	.179	.223	.404	.204	.523	.271	.608*	.336
Reach	.676*	.190	.179	.503	.327	.486	.304	.675*	.414
	.676*	.159	.187	.562	.325	.509	.301	.681*	.395
	.646*	.210	.220	na	.252	na	.312	.652*	.433
	na	na	na	na	na	na	na	na	na

Listed for each dimension:
  1st line: U.S. Army Flyers, male
  2nd line: U.S. Air Force Officers, male
  3rd line: U.S. Air Force Trainees, male
  4th line: U.S. Air Force Women
  5th line: U.S. Army Women

na: (data) not available
*: above 0.6
**: above 0.7

Table 1-17. (continued)

Design Dimensions	Stature	Bitragion Breadth	Head Length	Heel-Ankle Circumference	Hip Breadth, standing	Palm Length	Shoulder Circumference	Waist Height	Weight
Hip Breadth,	.331	.344	.208	.522	.891**	.182	.723**	.289	.832**
Sitting	.372	.312	.159	.512	.903**	.231	.669*	.295	.855**
	.360	.364	.202	.496	.921**	.162	.713**	.210	.875**
	.348	.266	.183	na	na	na	.606*	.318	.770**
	na	na	na	na	na	na	na	na	na
Chest	.198	.408	.229	.466	.706**	.162	.883**	.193	.878**
Circumference	.240	.398	.196	.466	.647*	.217	.822**	.203	.832**
	.246	.363	.214	.476	.721**	.172	.872**	.143	.861**
	.257	.309	.204	na	.551	na	.810**	.238	.799**
	.239	na	.176	na	na	na	na	na	na
Waist	.180	.377	.186	.430	.769**	.115	.778**	.171	.890**
Circumference	.224	.349	.158	.424	.724**	.154	.720**	.142	.856**
	.257	.377	.189	.447	.816**	.122	.764**	.117	.886**
	.279	.353	.149	na	.600	na	.775**	.238	.824**
	.208	na	.145	na	.571	na	na	.061	.787**
Hand	.441	.303	.267	.565	.300	.439	.351	.383	.423
Breadth	.409	.239	.209	.568	.329	.420	.378	.360	.448
	.390	.249	.259	.561	.348	.465	.452	.329	.485
	.380	.275	.241	na	.223	na	.355	.340	.417
	.433	na	.326	na	na	na	na	na	na
Head	.392	.258	.274	.612*	.378	.418	.440	.356	.510
Circumference	.412	.318	.231	.616*	.371	.422	.427	.365	.510
	.442	.279	.276	.626*	.387	.394	.502	.380	.539
	.365	.310	.236	na	.264	na	.424	.321	.495
	.448	na	.314	na	na	na	na	na	na
Hand Length	.661*	.189	.261	.572	.254	.859**	.250	.648*	.339
	.651*	.177	.246	.618*	.309	.844**	.258	.642*	.389
	.624*	.165	.243	.626*	.252	.857**	.288	.632*	.353
	.601*	.229	.316	na	.196	na	.289	.603*	.383
	na	na	na	na	na	na	na	na	na
Head Breadth	.065	.610*	.066	.176	.200	.052	.255	.035	.272
	.132	.622*	.058	.215	.227	.121	.244	.088	.305
	.088	.633*	.103	.190	.289	.036	.328	.016	.348
	.136	.586	.115	na	.188	na	.252	.122	.290
	.154	na	.161	.241	na	na	.302	.115	.317
Head	.265	.320	.694*	.330	.234	.175	.318	.202	.355
Circumference	.294	.424	.778**	.365	.288	.233	.327	.232	.412
	.258	.357	.809**	.371	.320	.190	.392	.182	.446
	.331	.408	.692*	na	.265	na	.330	.306	.403
	.364	na	.796**	.447	na	na	.352	.319	.409
Head Length	.286	.145	—	.327	.180	.206	.240	.221	.293
	.249	.126	—	.286	.166	.196	.201	.208	.261
	.262	.095	—	.334	.188	.205	.260	.216	.308
	.318	.158	—	na	.183	na	.248	.297	.304
	.370	na	—	.388	na	na	.246	.314	.317

Listed for each dimension:
  1st line: U.S. Army Flyers, male
  2nd line: U.S. Air Force Officers, male
  3rd line: U.S. Air Force Trainees, male
  4th line: U.S. Air Force Women
  5th line: U.S. Army Women

na: (data) not available
*: above 0.6
**: above 0.7

Glossary of Anatomical and Anthropometric Terms

abduct   To move away from the body or one of its parts; opposite of *adduct.*

acromion   A landmark on top of the shoulder: the highest point on the lateral edge of the scapula above the shoulder joint, at about half the width of the shoulder. Acromial height is usually equated with shoulder height.

adduct   To move toward the body; opposite of *abduct.*

anterior   In front of the body; toward the front of the body; opposed to *posterior.*

axilla   Armpit.

biceps brachii   The large muscle on the anterior surface of the upper arm, connecting the scapula with the radius.

biceps femoris   A large posterior muscle of the thigh.

brachialis   Muscle connecting the midhumerus with the ulna.

buttock protrusion   The maximal posterior protrusion of the right buttock.

carpus   The wristbones, collectively.

cervicale   Protrusion of the spinal column at the base of the neck caused by the tip of the spine of the 7th cervical vertebra.

clavicle   The "collarbone" linking the scapula with the sternum.

condyle   Articular prominence of a bone.

coronal plane   Any vertical plane at right angles to the midsagittal plane (same as *frontal plane*).

dactylion   The tip of the middle finger.

distal   The body segment (or section of it) farthest from the head (or the center of the body); opposite of *proximal.*

dorsal   Toward the back or spine; also pertaining to the top of a hand or foot, opposite of *palmar, plantar,* and *ventral.*

ear-eye line   An easily established reference line for the tilt angle of the head. It runs through the right meatus (ear hole) and the right external canthus (meeting corner of the eye lids). The EE line is angled about 11 deg. above the Frankfurt line.

epicondyle   Bony eminence at the distal end of the humerus, radius, and femur.

extend   To move adjacent segments so that the angle between them is increased, as when the leg is straightened; opposite of *flex.*

external   Away from the central long axis of the body; the outer portion of a body segment.

femur   The thigh bone.

flex   To move a joint in such a direction as to bring together the two parts that it connects, as when the elbow is bent; opposite of *extend.*

Frankfurt plane   The former standard horizontal plane for orientation of the head. The plane is established by a line passing through the right tragion (approximate ear hole) and the lowest point of the right orbit (eye socket), with both eyes on the same level. See ear-eye line.

frontal plane   Any vertical plane at right angles to the midsagittal plane (same as *coronal plane*).

glabella   The most anterior point of the forehead between the browridges in the midsagittal plane.

glenoid cavity   Depression in the scapula below the acromion, into which the head of the humerus fits, forming the shoulder joint.

gluteal furrow   Furrow at the juncture of the buttock and the thigh.

humerus   The bone of the upper arm.

iliac crest   The superior rim of the pelvic bone.

illium   See *pelvis.*

inferior   Below; lower in relation to another structure.

inseam   A term used in tailoring to indicate the inside length of a sleeve or trouser leg; it is measured on the medial side of the arm or leg.

internal   Near the central long axis of the body; the inner portion of a body segment.

ischium   The dorsal and posterior of the three principal bones that compose either half of the pelvis.

knuckle   The joint formed by the meeting of a finger bone (phalanx) with a palm bone (metacarpal).

lateral   Lying near or toward the sides of the body; opposite of *medial.*

malleolus   A rounded bony projection in the ankle region. The tibia has such a protrusion on its medial side, and the fibula one on its lateral side.

medial   Lying near or toward the midline of the body; opposite of *lateral.*

medial plane   The vertical plane which divides the body (in the anatomical position) into right and left halves; same as *midsagittal plane*

metacarpal   Pertaining to the long bones of the hand between the carpus and the phalanges.

midsagittal plane   The vertical plane that divides the body (in the anatomical position) into right and left halves; same as *medial plane.*

olecranon   Proximal end of the ulna.

omphalion   Center point of the navel.

orbit   Eye socket.

palmar   Pertaining to the palm (inside) of the hand; opposite of **dorsal.**

patella   Kneecap.

pelvis   The bones of the "pelvic girdle," consisting of the illium, pubic arch, and ischium, which compose either half of the pelvis.

phalanges   The bones of the fingers and toes (singular, *phalanx*).

plantar   Pertaining to the sole of the foot.

popliteal   Pertaining to the ligament behind the knee or to the part of the leg behind the knee.

posterior   Pertaining to the back of the body; opposite of *anterior.*

proximal   The (section of a) body segment nearest the head (or the center of the body); opposite of *distal.*

radius   The bone of the forearm on its thumb side.

sagittal   Pertaining to the medial (midsagittal) plane of the body or to a parallel plane.

scapula   Shoulder blade.

scye   A tailoring term to designate the armhole of a garment; refers here to landmarks that approximate the lower level of the axilla.

sphyrion   The most distal extension of the tibia on the medial side under the malleolus.

spine   The stack of vertebrae.

spine (or spinal process) of a vertebra   The posterior prominence.

sternum   The breastbone.

stylion   The most distal point on the styloid process of the radius.

styloid process   Long, spinelike projection of a bone.

sub-   A prefix designating below or under.

superior   Above, in relation to another structure; higher.

supra-   Prefix designating *above* or *on*.

tarsus   The collection of bones in the ankle joint.

tibia   Medial bone of the lower leg (shinbone).

tibiale   Uppermost point of the medial margin of the tibia.

tragion   The point located at the notch just above the tragus of the ear.

tragus   Conical eminence of the auricle (pinna, external ear) in the front wall of the ear hole.

transverse plane   Horizontal plane through the body, orthogonal to the medial and frontal planes.

triceps   The muscle of the posterior upper arm.

trochanterion   The tip of the bony lateral protrusion of the proximal end of the femur.

tuberosity   Large rounded prominence on a bone.

ulna   The bone of the forearm on its little-finger side.

umbilicus   Depression in the abdominal wall where the umbilical cord was attached to the embryo.

ventral   Pertaining to the anterior (abdominal) side of the trunk.

vertebra   A bone of the spine.

vertex   Top of the head.

Chapter 2

The Skeletal System

OVERVIEW

The skeletal system of the human body is composed of some 200 bones, of their articulations, and of connective tissue, all consisting of special cells embedded in an extracellular matrix of fibers in a ground substance. Bones provide the structural framework for the body. Ligaments connect the bones at their articulations, and tendons connect muscle with bone. The spinal column is often of great concern to the engineer because it is the locus of many overexertion injuries at work.

THE MODEL

Joints provide the links between long bones. In turn, the bones provide the basic framework for the body and the lever arms, to which tendons attach for muscle torque exertion about the joints. Mobility at the joints is limited by their design, by cartilage and other soft tissue, and by muscles.

CONNECTIVE TISSUE

Bones and other connective tissues are composed of cells embedded in a matrix of fibers and a ground substance. The cells in cartilage are called chondrocytes, while fibroblasts are found in loose tissue, such as skin, tendons, ligaments, and adipose tissue. The cells are enclosed by the extracellular matrix, which contains two different kinds of fibers: collagen fibers (subdivided into three types), which have high tensile strength and resist deformation, particularly stretch, and elastic fibers, which elongate. The ground substance has large molecules (proteoglycans) with a protein core, proteins (glycoproteins), calcium and other minerals (in bone), lipids, and water. The actual composition of tissues differs between bone, cartilage, muscles, and other connective tissues.

Bones

The main function of human skeletal bone (see Fig. 2-1) is to provide the internal framework for the whole body; without its support, the entire body would collapse into a heap of soft tissue. One distinguishes between flat, axial (appendicular) bones, such as those in the skull, sternum and ribs, and pelvis, and long, more

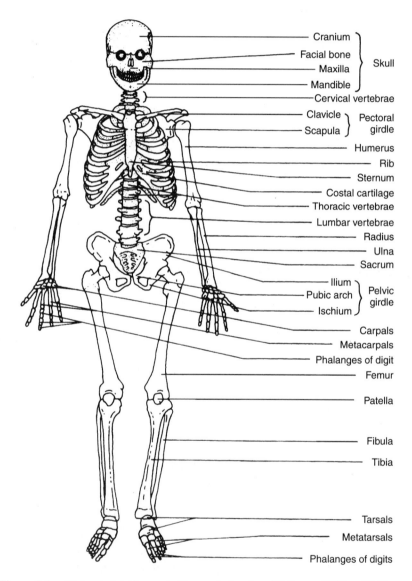

Figure 2-1.   Major skeletal bones (with permission from Weller and Wiley, 1979)

or less cylindrical bones. The long bones consist of a shaft (diaphysis) which, at each end, broadens into a metaphysis that forms part of the articulation to the adjacent bone. Both flat and long bones consist of compact (cortical) and spongy (cancellous) material. Cortical bone has a density of approximately 1.3 g/cm³ or more, while cancellous bone generally has a density of less than 1 g/cm³. This distinction between dense cancellous bone and porous cortical bone can be somewhat arbitrary, and sometimes the distinction is made depending on location rather than density. Cortical bone makes up the shaft (diaphysis) of long bones. Cancellous bone is found mostly at the ends (the epiphyses) where it helps to transfer the load experienced from an adjacent bone.

Furthermore, bones can act as shells to protect body portions; for example, the rib cage protects the lungs and heart. Long, hollow bones also provide the room for bone marrow, which serves as a blood factory. Finally, bones are a reservoir of calcium and phosphorus. Mechanically, bones are the lever arms at which muscles pull about the articulations (see Chap. 5).

Although bone is firm and hard and, thus, can resist deformation, it still has certain elastic properties. Bone is particularly compliant in childhood, when mineralization (ratio of the contents of inorganic material, mostly calcium, to organic substance) is relatively low, about 1:1. In contrast, the bones of the elderly are highly mineralized, about 7:1 and, therefore, tend to be more brittle.

Bone develops from a soft, woven-fibered material in earliest childhood into compact, mostly lamellar material with a hard outer shell and a spongy inner section. Growth takes place until about 30 years of age, with increasing elastic modulus (indicating the stiffness of the bone) and yield strength (indicating the nominal stress at which the bone undergoes a specified permanent deformation). After the third decade of life, bones usually become more and more osteoporotic, which means primarily a decrease in mass, due mostly to larger pores or holes, which decrease the structural strength of the whole bone. Osteoporotic bones also tend to have a reduction in the diaphyseal thickness accomplished by increases in the outer and inner diameters of long bones, bringing about a "hollowing" of their core; the remaining walls are often more brittle than at an earlier age because of an increase in mineral content in the existing matter (Ostlere and Gold, 1991).

An example of geometric changes in bone is adapted from Mow and Hayes (1991). Using measurements of the diaphysis of the human femur (thigh bone) from an archaeologic sample, parameters relevant to the strength of bone may be calculated. Increasing cross-sectional area indicates increasing axial strength, while increasing elastic section modulus indicates increasing bend strength. Area (A) and section modulus (I/c), respectively, are calculated using the formulae pertaining to a hollow tube with its outer and inner radii:

$$A = \pi \left(r_{outer}^{2} - r_{inner}^{2} \right)$$

$$I/c = \pi \left(r_{outer}^{4} - r_{inner}^{4} \right) / 4 / r_{outer}$$

The values are listed in Table 2-1.

Table 2-1. Effect of age-related dimensional changes in the human femur.

	Outer radius	Inner radius	Cross-sectional area	Section elastic modulus
Young adults	1.09 cm	0.53 cm	2.85 cm^2	0.96 cm^3
Elderly adults	1.25 cm	0.76 cm	3.09 cm^2	1.32 cm^3
Change with aging	+15%	+43%	+8%	+38%

Measurements on the femurs taken from young and elderly adults showed that both the outer and inner radii increased with aging, moving the outline of the bone outward. The net effect was only a small change in the cross-sectional area of the bone but a large change in the elastic section modulus associated with a large increase in the bending resistance of the bone. Since the bones of the elderly have reduced tensile strength, this geometric remodeling by hollowing helps maintain nearly constant strength of the whole bone.

Bone cells are nourished through canals carrying blood vessels and tubules. Bone is continuously resorbed and rebuilt throughout one's life; local strain encourages growth, while disuse leads to resorption (known as Wolff's law in the field of orthopedics). Adult bone material consists of cells in a fibrous, organic matrix (osteoid) consisting of collagen fibers and amorphous ground substance. Bone cells (osteocytes) have long, branching processes that occupy cavities (lacunae) in a matrix of densely packed collagenous fibers which, in turn, are in an amorphous ground substance (called cement) with a high-calcium phosphate content.

Cartilage

Another major component of the human skeletal system is cartilage, a translucent material of collagen fibers embedded in a binding substance. Cartilage supports moderate deformations and is firm but elastic, flexible, and capable of rapid growth. In the human, it supplies elastic structures where required, such as between bones to allow or restrict movement and at the ends of bones to define flexible structures such as the external ear and the nose.

Cartilage is divided into three types: hyaline cartilage, elastic cartilage, and fibrocartilage. Hyaline cartilage is smooth, glistening, and most commonly found at the articulating surfaces of long bones, such as at the knee joint (see "Articulations," later in this chapter). It is also found in the nasal septum, the larynx (voice box), and at the ribs. Elastic cartilage is generally more flexible than hyaline and is found at the external auditory canal and the eustachian tube. Fibrocartilage makes up the annulus fibrosus of the intervertebral disk (see "The Disk" later in this chapter) and the meniscus in the knee (Mow and Hayes, 1991).

Other Connective Tissues

Dense connective tissues, mostly composed of collagen fibers, are referred to as ligaments when they connect bones, as tendons when they connect muscle with bone, and as fascia when they wrap organs or muscles. The wrapping fascial tissue of muscle fibers condenses at the ends of the muscle to tendons, which are usually encapsulated by a fibrous tissue called sheaths. These allow a gliding motion of the tendon against surrounding materials through an inner lining, or synovium, which produces a viscous fluid, synovia, that reduces friction. Particularly at the wrist and fingers, rings or bands of ligaments keep the tendon sheaths close to underlying bones, acting as guides or pulleys for the pulling actions of the tendons and their muscles.

ARTICULATIONS

In adults, some bony joints, such as the seams in the skull, have no mobility left; some, such as the connections of the ribs to the sternum, have very limited mobility. Joints with one degree of freedom are simple hinge joints, like the elbow or the distal joints of the fingers. Other joints have two degrees of freedom, such as the ill-defined wrist joint, where the hand may be bent in flexion and extension and laterally pivoted. (The capability to twist is located in the forearm, not in the wrist.) Other joints, such as shoulder and hip joints, have three degrees of freedom.

Synovial fluid in a joint facilitates movement of the adjoining bones by providing lubrication. For example, while a person is running, the cartilage in the knee joint can show an increase in thickness of about 10%, brought about in a short time by synovial fluid seeping into it from the underlying bone marrow cavity. Similarly, fluid seeps into the spinal disks when they are not compressed, for example, during sleep. This makes the disks more pliable directly after one gets up than during the day, when the fluid is "squeezed out" by the load of body masses and their accelerations. Thus, immediately after getting up, one stands taller than after a day's effort.

Bones are connected by articulations that may be considered the bearing surfaces of adjacent bones. The design of the bones at their joints, the encapsulation by ligaments, the supply of cartilaginous membranes, and the provision of disks or volar plates, together with the action of muscles, determine the mobility of body joints. Depending on their structure, these joints allow no, little, or much relative displacement between the adjacent bones. Figure 2-2 shows different types of joints.

Fibrous joints are junctions of bones that are fastened together by fibrous tissue; no tissue separates the bone surfaces. Fibrous joints, such as those in the skull allow no appreciable motion. *Cartilaginous* joints provide very limited movement: here, articular cartilage lines the opposing ends of the bones, and a tight ligament covers the joint and extends along both bone endings. The spinal column has cartilaginous joints, in which a flat disk of fibrocartilage connects the linings of two opposing vertebral bodies. In *synovial* joints, the opposing bones are also lined by articular cartilage at the opposing bone surfaces, but these are separated by a space. The synovial joint is loosely encapsulated by an elastic ligament, allowing much relative displacement of the bones. In some synovial joints, a fibrocartilage disk or wedge (such as the meniscus in the knee) is in the joint space, allowing high mobility. For further information on specific joints, see, for example, Frankel and Nordin (1980).

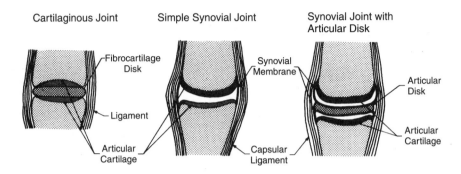

Figure 2-2. Types of body joints (modified from Astrand and Rodahl, 1977, and from Chaffin and Andersson, 1991).

Articular cartilage has no blood vessels. Its nourishment is achieved through direct communication between the cavities of the end portion of the bone (epiphyses) and the basal portions of the articular cartilage. Also, synovial fluid may be exchanged between cartilage and articular space. Synovial membranes can secrete synovia, which acts as a lubricant. A strained joint, such as the knee joint following a running injury, can show an increase in cartilage thickness of 10% or more, brought about in a few minutes by fluid seeping into the cartilage from the underlying bone marrow cavity. This helps to "fill and smooth out" the space between the opposing bones of the joint, reducing the danger of local pressures and damage. Similarly, fluid seeps into the spinal disks when they are not compressed, such as during a night's bed rest. This makes the disks more pliable, which may explain disk deformations experienced as early morning back pains upon waking and rising.

Mobility

Most articulations of the human body (excepting the spine) belong to the synovial group. Movements in such joints range from a simple gliding movement to rather complex movements in several planes. Examples are the rotatory pronation and supination of the hand achieved by relative gliding-twisting of the radius and ulna within the forearm (one degree of freedom), the hinge-joint type angular movement of the forearm around the upper arm in the elbow joint (also one degree of freedom), or the complex spatial circumduction that can be performed (with three degrees of freedom) in the shoulder or hip. The motions themselves are limited by the shapes of the bony surfaces and by the restraining tension generated by ligaments and muscles.

Mobility, less correctly also called flexibility, indicates the range of motion that can be achieved at body articulations. It is usually measured as the difference between the smallest and largest angles enclosed by neighboring body segments about their common point of articulation. Figure 2-3 illustrates possible motions

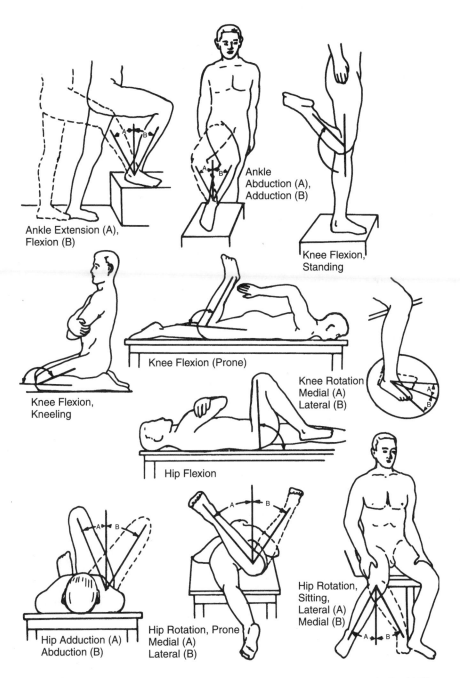

Figure 2-3. Maximal displacement at body joints (from Van Cott and Kinkade, 1972).

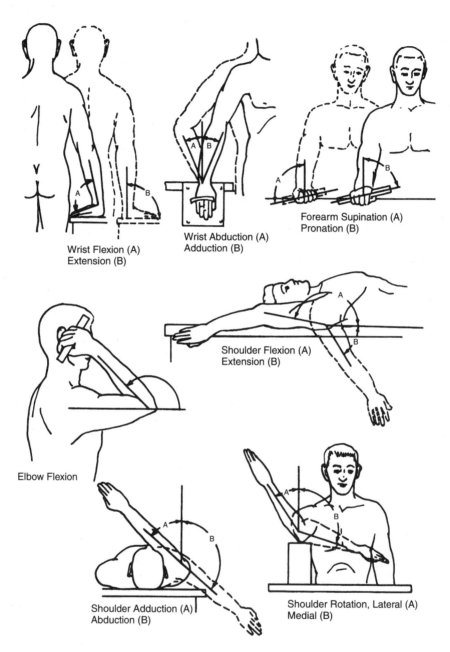

Wrist Flexion (A)
Extension (B)

Wrist Abduction (A)
Adduction (B)

Forearm Supination (A)
Pronation (B)

Shoulder Flexion (A)
Extension (B)

Elbow Flexion

Shoulder Adduction (A)
Abduction (B)

Shoulder Rotation, Lateral (A)
Medial (B)

Figure 2-3.   (Continued)

at major joints. Unfortunately, these data are not as clear and clean as they appear: quite often, the actual point of rotation moves with the motion. For example, the geometric location of the axis of the knee joint moves slightly while the thigh and the lower leg rotate about it. Furthermore, the actual arms of the rotation angle are not well defined: they may be established by straight lines between the center of rotation of the enclosed articulation to the point of rotation of the next articulation (in the example of the knee joint, running distally to the ankle, and proximally to the center of the hip joint); or they may simply be estimated as representing the midaxes of the adjacent body and/or bone segments.

The range of motion varies with gender, training and physical condition, and age. It depends on whether only active contraction of the relevant muscles is used or whether gravity, or perhaps even a second person, helps to achieve extreme locations. Finally, of course, the mobility is different in different planes if the articulation in question has more than one degree of freedom.

A study by Staff (1983) has provided reliable information about voluntary (unforced) mobility in major body joints. This study was done on 100 females (ages 18–35) and carefully controlled to resemble an earlier study by Houy (1983) on 100 male subjects. On each person, only one measurement was taken for each motion, recorded by an electronic bubble goniometer. All subjects were asked to move their limbs "only as far as comfortably possible," using (if applicable) the dominant limb; 85% claimed to be right-dominant. The results are compiled in Table 2-2. Of the 32 measurements, 24 showed significantly more mobility by females than by males; men had larger mobility only in ankle flexion and wrist abduction. This finding confirms earlier studies that also showed larger motion capability, in most cases, in women. Given the careful and consistent way the two studies were conducted, one may assume that the data in Table 2-3 are a true reflection of the adult U.S. population within the working age span. Then, it seems that mobility is only slightly reduced (less than about 10 deg in the extreme positions) for most body joints as one reaches the sixth decade.

Control of motions in the joints is mostly effected by the muscles that span them. Movable joints have nervous connections with the muscles that act on them, establishing local reflex arcs that prevent overextensions. Four types of nerve endings exist in the joints. Three of them terminate in specialized organs, called Ruffini organs. They provide information about changes in joint position, speed of movement, and actual positioning of the joint. The two first Ruffini organs are located in the joint capsule; the other one is in the ligament. The fourth receptor is a free branching nerve ending in pain-sensitive fibers. Synovial membranes and joint cartilage do not have nerve receptors.

Artificial Joints

Natural joints may fail as a result of disease, trauma, or long-term wear and tear. The degeneration of cartilage in the major joints (such as the hip or knee) may lead to the replacement of the articulating surfaces with artificial, manufactured joints if "conservative" medical treatment fails. This is routinely done in hips and knees as well as fingers. In the United States annually, more than 400,000 joints are implanted, predominantly in elderly persons. Although total joint replacement

Table 2-2. Comparison of mobility data for females and males in degrees (adapted from Staff 1983).

Joint	Movement	5th percentile		50 percentile		95th percentile		Difference*
		Female	Male	Female	Male	Female	Male	Female > Male
Neck	Ventral Flexion	34.0	25.0	51.5	43.0	69.0	60.0	+8.5
	Dorsal Flexion	47.5	38.0	70.5	56.5	93.5	74.0	+14.0
	Right Rotation	67.0	56.0	81.0	74.0	95.0	85.0	+7.0
	Left Rotation	64.0	67.5	77.0	77.0	90.0	85.0	NS
Shoulder	Flexion	169.5	161.0	184.5	178.0	199.5	193.5	+6.5
	Extension	47.0	41.5	66.0	57.5	85.0	76.0	+8.5
	Adduction	37.5	36.0	52.5	50.5	67.5	63.0	NS
	Abduction	106.0	106.0	122.5	123.5	139.0	140.0	NS
	Medial Rotation	94.0	68.5	110.5	95.0	127.0	114.0	+15.5
	Lateral Rotation	19.5	16.0	37.0	31.5	54.5	46.0	+5.5
Elbow-Forearm	Flexion	135.5	122.5	148.0	138.0	160.5	150.0	+10.0
	Supination	87.0	86.0	108.5	107.5	130.0	135.0	NS
	Pronation	63.0	42.5	81.0	65.0	99.0	86.5	+16.0
Wrist	Extension	56.5	47.0	72.0	62.0	87.5	76.0	+10.0
	Flexion	53.5	50.5	71.5	67.5	89.5	85.0	+4.0
	Adduction	16.5	14.0	26.5	22.0	36.5	30.0	+4.5
	Abduction	19.0	22.0	28.0	30.5	37.0	40.0	-2.5
Hip	Flexion	103.0	95.0	125.0	109.5	147.0	130.0	+15.5
	Adduction	27.0	15.5	38.5	26.0	50.0	39.0	+12.5
	Abduction	47.0	38.0	66.0	59.0	85.0	81.0	+7.0
	Medial Rotation (Prone)	30.5	30.0	44.5	46.0	58.5	62.5	NS
	Lateral Rotation (Prone)	29.0	21.5	45.5	33.0	62.0	46.0	+12.5
	Medial Rotation (Sitting)	20.5	18.0	32.0	28.0	43.5	43.0	+4.0
	Lateral Rotation (Sitting)	20.5	18.0	33.0	26.5	45.5	37.0	+6.5
Knee	Flexion (Standing)	99.5	87.0	113.5	103.5	127.5	122.0	+10.0
	Flexion (Prone)	116.0	99.5	130.0	117.0	144.0	130.0	+13.0
	Medial Rotation	18.5	14.5	31.5	23.0	44.5	35.0	+8.5
	Lateral Rotation	28.5	21.0	43.5	33.5	58.5	48.0	+10.0
Ankle	Flexion	13.0	18.0	23.0	29.0	33.0	34.0	-6.0
	Extension	30.5	21.0	41.0	35.5	51.5	51.5	+5.5
	Adduction	13.0	15.0	23.5	25.0	34.0	38.0	NS
	Abduction	11.5	11.0	24.0	19.0	36.5	30.0	+5.0

*Listed are only differences at the 50th percentile, and if significant ($\alpha < 0.5$)

typically restores function and mobility to the patient, its primary and most appreciated purpose is to relieve pain.

For the patient, joint degeneration is associated with pain and with progressive and severe limitations of motion. If needed, the articulating surfaces are typically replaced in their entirety: in the hip, the head of the femur (thigh bone) is removed and replaced by a spherical metallic ball on a stem, and the acetabular cup is resurfaced with a plastic liner. In the knee, the articulating surfaces on the bottom of the femur are replaced with metal, and articulating surfaces at the top of the tibia (shinbone) and on the patella (kneecap) are resurfaced with plastic. These joints all have the same type of design for the major load-bearing components: the metallic component is convex and the plastic component is concave. The plastic now used (after an early, disastrous attempt with Teflon) is an ultra-high-molecular-weight polyethylene.

Replacements for the ball-and-socket joint at the hip have been attempted for about a century. Routinely successful total hip replacement started with Sir John Charnley's work in England in the 1960s. He pioneered the use of the metal-on-plastic articulations and the use of PMMA (polymethyl methacrylate) as a bone "cement." This cement is used as a grout to link the prosthesis and the bone mechanically and has no adhesive properties. Today, at least 90% of patients with hip and knee replacements are pleased with their new joints, which function well 10 years and longer after surgery. Since surgical technique and device design have improved over the last decade, it is expected that today's joint-replacement patients will enjoy even higher success rates.

Recently, devices have been designed that attempt to fix the bone directly to the metal implant surfaces, avoiding the need for bone cement. These implant surfaces may be coated with small beads or thin wires to create a pore size of less than 1 mm into which the bone is supposed to grow. To encourage bony ingrowth, an osteoinductive or osteoconductive chemical coating may be sprayed on the "porous" surface of the implant.

If necessary, finger joints are usually replaced by a one-component, molded-plastic integral hinge. This simple artificial joint is successful for several reasons, including the low loads carried by the joints, and the minimal debris generated by wear.

The design of joint replacements is constrained by biologic and mechanical considerations. Biologically, the device must be compatible with the body, both in toto and in particulate form (such as wear debris). The material from which the device is made must be analyzed for toxicity, reactivity, and strength, particularly in such a corrosive and warm environment as the human body. The interaction between materials and rate of wear with and without lubrication must also be ascertained. The device must be implantable (in terms of complexity and size) and should yield near-normal range of motion. Finally, the design of the device should consider the possibility of salvage: sufficient bone and soft tissue should remain to allow for replacement of the device or fusion of the joint if needed.

When joint replacements fail, the symptom to the patient is usually severe pain. Upon examination, infection is commonly found and, often, the device is no longer firmly attached to the surrounding bone. This is mostly a mechanical problem, frequently associated with debris (wear particles) from the metallic component, the plastic component, or the cement or some combination of these. The wear particles may trigger a biologic response that leads to resorption of the bone and loss of implant support, as well as inflammation, reduced range of motion, and pain (Dumbleton, 1988; Elbert, 1991; Galante et al., 1991).

The Spinal Column

The spinal column is a particularly interesting and complex structure. As shown in Figure 2-4, it consists of 25 bony elements: 7 cervical, 12 thoracic, and 5 lumbar vertebrae, and the sacrum (with the coccyx) which consist of fused groups of rudimentary bones. These components are held together in cartilaginous joints. The main bodies of the vertebrae rest on each other on fibrocartilage spinal disks. Vertebrae also have two protuberances extending backward-upward; these superior articulation processes (Figs. 2-5 and 2-6) end in rounded surfaces fitting into cavities on the underside of the next-higher vertebra. These facet joints are covered with synovial tissue. The facet joints mostly determine the ability to twist the spine, which occurs chiefly in the cervical and thoracic sections. Hence, the vertebrae rest on each other in three joints of two different kinds: two synovial facet joints and the cartilaginous intervertebral body joint. This complex rod sustaining the trunk is kept in delicate balance by ligaments and particularly by muscles running along the posterior side of the spinal column and located along the sides and front of the trunk.

Figure 2-6 shows a top view of a typical vertebra: the main body to which the disk attaches; the archlike structure around the foramen that provides a protective passage for the spinal cord; the spinous (posterior) process and the two transverse (lateral) processes that provide attachment and leverage for muscles and ligaments; and the two superior processes that end in the facet joints. (The geometry of vertebrae has been described in exact detail by Panjabi et al., 1992).

Figure 2-5 is a schematic representation of the lumbar section of the spinal column showing particularly the bearing surfaces at the main bodies and at the facets. The spine is capable of withstanding considerable loads, yet is flexible enough to allow a large range of postures. There is, however, a trade-off between load carried and flexibility. If there is no external load on the spine, only its anatomical structures (bone geometry, joints, ligaments, and muscles) restrict its mobility. Applying load to the spinal column reduces its mobility until, under heavy load, the range of possible postures is very limited. The traditional model of the spine has been that of a straight column, as depicted in Figure 2-7A. This simplification allows a unique description of its geometry and strain under the applied load (Aspden, 1988; Yettram and Jackman, 1981). If we consider the spinal column as arches, as shown in Figure 2-7B and C, its load-bearing mechanism depends on the curvature. Since the geometry of spinal arching is not fixed, there is no unique solution that describes its strain. Force along the arch is thought to be transmitted along a straight line, called the thrust line. The theorem of plasticity assumes that, if the arch is to be stable, the thrust line must lie within the cross section of the arch components throughout its entire length. If, at any point, the

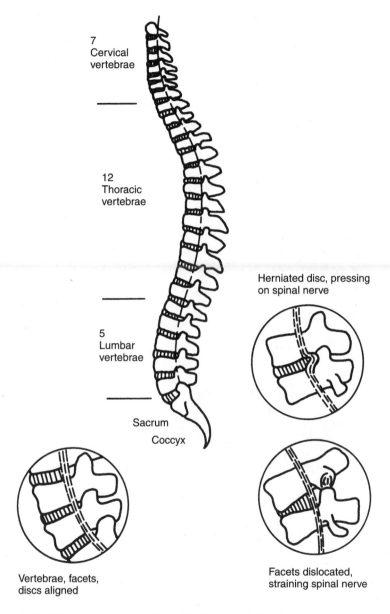

Figure 2-4. Schematic representation of the human spinal column.

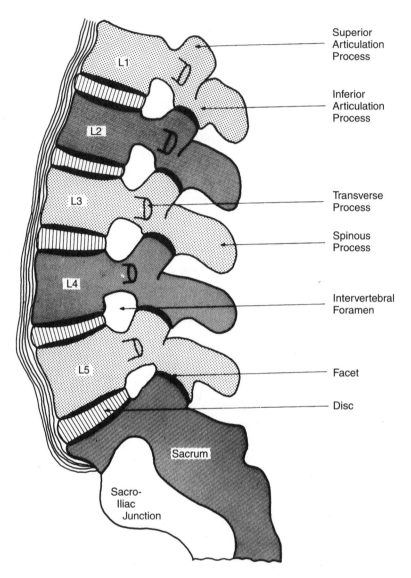

Figure 2-5. Schematic representation of the lumbar section of the spinal column (bearing surfaces are indicated in heavy lines).

thrust line lies outside the arch, tensile force must keep the arch within its possible position range or it buckles (see Fig. 2-7).

A major load on the spine is compression. Figure 2-8 illustrates that the compressive force results from the pull of trunk muscles and from the weight due to segment masses and external load. Spinal compression is somewhat relieved by

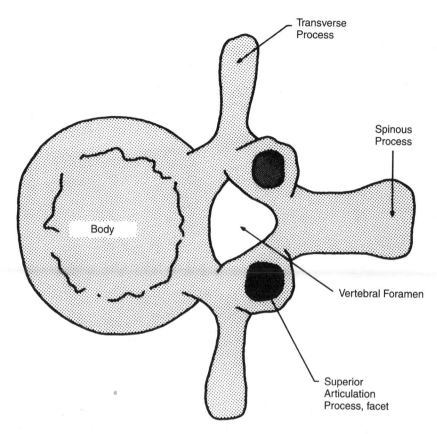

Figure 2-6.   Top view of a vertebra.

the upward-directed force due to intra-abdominal pressure (IAP). Yet, largely because of the slanted arrangement of load-bearing surfaces at disks and facet
joints, the spine is also subjected to shear. Furthermore, the spine must withstand
both bending and twisting torques. Aspden (1988) calculated spinal strain according to his model and obtained three interesting results:

1. The calculated compression loads in a stable arched spine are considerably lower than those computed using the straight model.
2. These loads depend on the adopted posture, that is, the geometry of the
   spine.
3. Intra-abdominal pressure (IAP) can stiffen the lumbar spine.

The effect of IAP on stiffening the lumbar spine is shown in Figure 2-7. The
thrust line would be outside the spinal column if kyphotic flattening of the lumbar area were maintained. Yet, if lumbar lordosis is introduced, the thrust line can

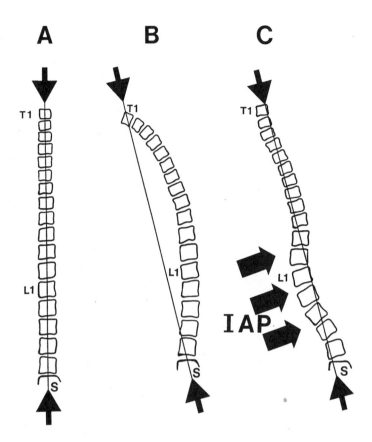

Figure 2-7. Models of the spinal column: (A) straight, (B) arched, (C) supported by intra-abdominal pressure (IAP). Modified from Aspden, 1988. With permission from Kroemer, K.H.E., Kroemer, H.B., and Kroemer-Elbert, K.E. (1994) *Ergonomics. How to Design for Ease and Efficiency.* Englewood Cliffs, NJ: Prentice-Hall. All rights reserved.

be kept within the spinal components, and the arch is stable. The larger the IAP, the better lordosis can be maintained, even under heavy axial loading (compression) of the spinal column. Such lordotic curvature of the spinal column is, supposedly, used by competitive weight lifters so that they can lift large weights with relatively small compressive force in the spine. In contrast, the straight spinal column (as presumed in traditional spinal modeling) generates large compression forces for the same external load (Aspden, 1988).

The spinal column is often the location of discomfort, pain, and injury because it transmits many internal and external strains. For example, when we are standing or sitting, impacts and vibrations from the lower body are transmitted primarily through the spinal column into the upper body. Conversely, forces and

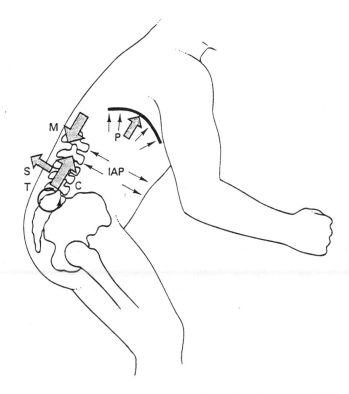

Figure 2-8.   Intra-abdominal pressure (IAP) and its resulting force vector (*P*) reduce the compressive force (*C*) produced by trunk muscle pull (*M*). Shear force (*S*) and torque (*T*) also load the spine. With permission from Kroemer, K.H.E., Kroemer, H.B., and Kroemer-Elbert, K.E. (1994) *Ergonomics. How to Design for Ease and Efficiency.* Englewood Cliffs, NJ: Prentice-Hall. All rights reserved.

impacts experienced through the upper body, particularly when we are working with our hands, are transmitted downward through the spinal column to the floor or seat structure that supports the body. Thus, the spinal column must absorb and dissipate much energy, whether it is transmitted to the body from the outside or generated inside by muscles for exertion of work to the outside.

The Disk

The nucleus pulposus, at the center of the spinal disk, has no blood supply or sensory nerves of its own. It is nourished through the disk as a result of osmotic pressure, gravitational force, and the pumping effects of spine movement. The nucleus is similar to a gel-filled deformable ball that allows the opposing surfaces of the vertebral bodies to tilt with respect to each other; it carries much of the compressive loading transmitted along the spine. The nucleus is held in position by layers of elastic tissues, the annulus fibrosus. Core and surrounding tissues are a physiological shock absorber. If the disk does not function well, for instance, as a result of injury or deterioration, it cannot properly transmit loads and allow motions between adjacent vertebrae. The situation is even worse when parts of the annulus fibers get out of place, especially when weakened or broken; this is often called "herniated disk." If matter protrudes toward the spinal cord or the nerve roots emanating from it, they may impinge on nerve tissue and affect the transmission of nerve signals. This, in turn, may be interpreted by the brain as a disorder in the body part from which that nerve reports, or it may hinder the nervous control of the body part. Typical of this is sciatica, an affliction of the sciatic nerve that traverses hip and thigh. Malfunction of the disk can bring about a combination of misalignment of vertebrae and strain of vertebral bone, of facet joints, and of connective tissues, muscles, and ligaments. It often leads to restricted mobility, reduced ability to perform physical work, particularly to move loads such as in lifting. Low-back pain is a common complaint that is often difficult to diagnose and to treat. It is the result of an inherent structural weakness, of an injury, or of diseases and is a condition that has plagued humans since ancient times. It was diagnosed among Egyptians 5,000 years ago and was discussed in 1713 by Bernadino Ramazzini. The problem is not confined to human beings since quadrupeds may suffer from low-back pain as well.

Aging affects the spine as it does all skeletal components. Bone mass and structural properties increase with age in the young, and then begin to decrease in the adult. Younger spines are up to 10 times more flexible than adult spines, depending on the load direction. Aging and growth also increase the angulation of the cervical facets, as well as cervical lordosis (concave-forward curvature) in adults. In the elderly, increases in thoracic kyphosis (convex-backward curvature) and decreased lumbar lordosis are seen; this is due, at least in part, to decreases in the height of the disk. Disk height changes are caused by degeneration of the disk and changes in the curvature of the vertebral end plate, possibly from osteoporosis (Mow and Hayes, 1991).

As already mentioned, the spinal column is very often the site of overexertion. This may manifest itself by muscular and cartilaginous strains and sprains or in deformation of, or even damage to, the disks or the vertebrae. Because such injuries constitute a frequent and costly problem in U.S. industries, much research has been directed at the causes and mechanisms involved (see, e.g., White and Panjabi, 1978). Since no detailed discussion of problems is possible here (or in Chap. 5), it must suffice to state that many problems seem to be associated with the muscles or cartilage stabilizing the bony stack of vertebrae. Uncoordinated pull of the muscles on this column, particularly when associated with high or asymmetric external loads, or both, may—directly or indirectly, acutely or cumulatively, often in unknown and not reconstructable ways—lead to various back strains and back injuries (Pope et al., 1984; Nordin et al., 1997). Some, but not all, occasions for such overexertion injuries can be avoided by ergonomic design of workplace, work equipment, and work task (Kroemer, 1997). Even for persons with persistent back pains, workplaces and procedures can be engineered to allow performance of suitable physical work (Kroemer, Kroemer, and Kroemer-Elbert, 1994; Rodgers, 1985).

SUMMARY

The skeletal system with its bones, joints, and connective tissue is, from the engineering point of view, highly complex. It allows a varying but well-controlled range of motion, even under high external loads. Its capabilities are highly variable: they are reduced with age but increased by loading, which strengthens strained bones and lubricates loaded joints. Control of the skeletal system is accomplished by nervous reflex feedback and by excitation signals from the central nervous system to muscles that stabilize and load bones and joints. The spinal column, in particular, has been, and continues to be, the object of many biomechanical and ergonomic studies.

REFERENCES

Aspden, R.M. (1988). A New Mathematical Model of the Spine and Its Relationship to Spinal Loading in the Workplace. *Applied Ergonomics,* 19:319–323.

Astrand P.O., and Rodahl, K. (1977). *Textbook of Work Physiology,* 2nd ed. New York, NY: Wiley.

Chaffin, D.B., and Andersson, G.B.J. (1991). *Occupational Biomechanics.* (2nd ed.) New York, NY: Wiley.

Dumbleton, J.H. (1988). The Clinical Significance of Wear in Total Hip and Knee Prostheses. *Journal of Biomaterials Applications,* 3:3–32.

Elbert, K.E.K. (1991). *Analysis of Polyethylene in Total Joint Replacement.* Doctoral dissertation, Cornell University, Ithaca, NY.

Houy, D.A. (1983). Range of Joint Motion in College Males. In *Proceedings of the Human Factors Society 27th Annual Meeting* (pp. 374–378). Santa Monica, CA: Human Factors Society.

Galante, J.O., Lemons, J., Spector, M., Wilson, P., and Wright, T.M. (1991). The Biologic Effects of Implant Materials. *Journal of Orthopedic Research,* 9:760–775.

Kroemer, K.H.E. (1997). *Ergonomic Design of Material Handling Systems.* New York: NY: CRC Press/Lewis Publishers.

Kroemer, K.H.E., Kroemer, H.B., and Kroemer-Elbert, K.E. (1994). *Ergonomics: How to Design for Ease and Efficiency.* Englewood Cliffs, NJ: Prentice-Hall.

Mow, V.C., and Hayes, W.C. (eds.) (1991). *Basic Orthopaedic Biomechanics.* New York, NY: Raven Press.

Nordin, M., Andersson, G.B.J., and Pope, M.H. (eds.) (1997). Work-Related Musculoskeletal Disorders. New York, NY: Mosby-Yearbook.

Ostlere, S.J., and Gold, R.H. (1991). Osteoporosis and Bone Density Measurement Methods. *Clinical Orthopaedics* 271:149–163.

Panjabi, M.M., Goel, V., Oxland, T., Takata, K., Duranceau, J., Krag, M., and Price, M. (1992). Human Lumbar Vertebrae: Quantitative Three-Dimensional Anatomy. *Spine* 17:299–306.

Pope, M.H., Frymoyer, J.W., and Andersson, G. (eds.) (1984). *Occupational Low Back Pain.* Philadelphia, PA: Praeger.

Rodgers, S.H. (1985). *Working with Backache.* Fairport, NY: Perinton.

Staff, K.R. (1983). *A Comparison of Range of Joint Mobility in College Females and Males.* Master's Thesis, Industrial Engineering. College Station, TX: Texas A&M University.

Van Cott, H.P., and Kinkade, R.G. (1972). *Human Engineering Guide to Equipment Design.* Washington, DC: U.S. Government Printing Office.

Weller, H., and Wiley, R.L. (1979). *Basic Human Physiology.* New York, NY: Van Nostrand Reinhold.

White, A.A., and Punjabi, M.M. (1978). *The Clinical Biomechanics of the Spine.* Philadelphia, PA: Lippincott.

Yettram, A.L., and Jackman, J. (1981). Equilibrium Analysis for the Forces in the Human Spinal Column and Its Musculature. *Spine* 5:402–411.

FURTHER READING

Burstein, A.H., and Wright, T.M. (1994). *Fundamentals of Orthopaedic Biomechanics.* Baltimore, MD: Williams & Wilkins.

Chaffin, D.B., and Andersson, G.B.J. (1991). *Occupational Biomechanics.* (2nd ed.) New York: NY: Wiley.

Currey, J. (1984). *The Mechanical Adaptations of Bone.* Princeton, NJ: Princeton University Press.

Frankel, V.H., and Nordin, M. (1980). *Basic Biomechanics of the Skeletal System.* Philadelphia, PA: Lea & Febiger.

Fung, Y.C. (1993). *Biomechanics: Mechanical Properties of Living Tissues,* 2nd ed. New York, NY: Springer-Verlag.

Mow, V.C., and Hayes, W.C. (eds) (1991). *Basic Orthopaedic Biomechanics.* New York, NY: Raven Press.

Rodgers, S.H. (1985). *Working with Backache.* Fairport, NY: Perinton.

White, A.A. (1985). *Your Aching Back.* Toronto: Bantam.

Winter, D.A. (1990). *Biomechanics and Motor Control of Human Movement,* 2nd ed. New York, NY: Wiley.

Chapter 3

Skeletal Muscle

OVERVIEW

Skeletal muscles move body segments with respect to each other against internal and external resistances. Shortening is the only active function of the muscle. Active shortening, or contraction, of a muscle is controlled in the central nervous system by rate (also called recruitment) coding. These signals stimulate muscle components to shorten dynamically, to retain their length statically, or to permit passive lengthening. Various methods and techniques are available for assessing muscular control and strength. The engineering application of data on available body strength requires the determination of whether minimal or maximal exertions, either static or dynamic, are the critical design considerations.

THE MODEL

The human body is commonly modeled from the inside out: skeletal bones form an internal framework, which is movable in its intermediate articulations. Shapes, volumes, and mass properties are attached to the structural segments. Muscles with their tendons are attached at origin and insertion ends to bones, spanning one or two body joints.

Muscles generate forces that pull on the bones, which act as lever arms. The resulting torques about the body joints are regulated by nervous control.

BACKGROUND AND TERMINOLOGY

Muscular efforts have been of special interest to physiological science; therefore, there is a long tradition of philosophical and experimental approaches and use of terminology.

Leonardo da Vinci (1452–1519) and Giovanni Alfonso Borelli (1608–1679) combined mechanical with anatomical and physiological explanations to describe the functioning of the human body. Since Borelli, the human body has usually been modeled as consisting of long bones (links) that are connected in the articulations (joints), powered by muscles that bridge the articulations. The knowledge developed by Leibnitz (1646–1716) and Newton (1642–1727) explained the physical relationships between force, mass, and motion. Of particular importance are Newton's three laws:

- The first law explains that unbalanced force acting on a mass changes its motion condition (that is, accelerates or decelerates a mass).
- The second law states that force equals mass multiplied by acceleration; $F = m * a.$
- The third law makes it clear that force exertion requires the presence of an equally large counterforce.

Physiology books published until the middle of the 20th century tended to divide muscle activities either into dynamic efforts lasting for minutes or hours, with work, energy, and endurance typical topics, or into short bursts of contraction force exertion. Research on muscle effort concentrated on the *isometric* condition. Consequently, most information on muscle strength was gathered for such static exertion. All other muscle activities were typically called *anisometric,* often falsely labeled "isotonic" or "kinetic," meant to cover all the many possible dynamic muscle uses. Appendix A at the end of this chapter lists and defines terms that correctly describe muscular events.

ARCHITECTURE

The human body has three types of muscle: skeletal (striated), smooth, and cardiac muscles. Skeletal muscles (also called voluntary muscles) control body locomotion and posture. Smooth muscles control the opening of, and flow in, blood vessels and internal organs. Cardiac muscle in the heart pumps blood through the vascular system. Electrical stimulation and contraction mechanism are similar in smooth and cardiac muscle, while anatomic and physiologic characteristics of cardiac muscle resemble those of skeletal muscle.

For the engineer, skeletal muscles are of primary interest because they move the segments of the human body and generate energy for exertion against outside objects. Skeletal muscles connect two body links across their joint, or even three links across two intermediate joints.

Agonist-Antagonist, Cocontraction

In the human body, muscles are usually arranged in functional pairs so that a contracting muscle is counteracted by its opponent. One muscle, or one group of synergistic muscles, flexes while the other extends, as shown in Figure 3-1. The active muscle is named *agonist* (also called protagonist), and its opposite is named *antagonist. Cocontraction*—the simultaneous contraction of two or more muscles, often of agonist and antagonist—generally serves to control the magnitude of a

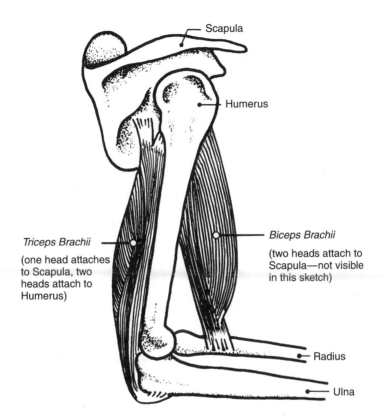

Figure 3-1. Biceps and triceps muscles as an antagonistic pair controlling elbow flexion and extension. Not shown are the brachialis muscle (attached to the humerus and ulna) and the brachioradialis muscle (connecting humerus and radius), which act together with the biceps as a synergistic flexor group.

strength exertion or the speed of the motion of the limbs. Another kind of cocontraction occurs when muscles activate that are not directly involved in a task. This happens, for example, in tightening muscles in the left arm when those in the right arm execute a strong effort; this is called *bilateral cocontraction.*

COMPONENTS OF MUSCLE

There are several hundred skeletal muscles in the human body, known by their Latin names. They actually consist of bundles of muscles (fasciculi), each of which is wrapped, as is the total muscle, in connective tissue (fascia), which embeds nerves and blood vessels. The sheaths of this connective tissue influence the mechanical properties of muscle. At the ends of the muscle, the tissues combine to form tendons, which attach the muscle to bones. By weight, muscle consists of

75% water, 20% protein, and 5% other constituents, such as fats, glucose and glycogen, pigments, enzymes, and salts. The structural elements of skeletal muscles are listed in Table 3-1.

Thousands of individual muscle fibers run, more or less parallel, the length of the muscle. These fibers are enveloped by a membrane (sarcolemma). Inside, they contain sarcoplasm and up to several hundred nuclei, particularly mitochondria. The mitochondrion is the power factory of the cell, providing and converting chemically stored energy, in ATP (adenosine triphosphate), for the physical work of muscle contraction (see Chap. 8 for details).

Seen via a microscope, skeletal muscle fibers appear striped (striated): thin and thick, light and dark bands run across the fiber in regular patterns, which repeat each other along the length of the fiber. One such dark stripe appears to penetrate the fiber like a thick membrane or disk: this is the so-called z-disk (from the German *zwischen*, "between"), which carries the "plumbing and control" networks through the muscle tissue, as discussed in the following subsection. The distance between two adjacent z-lines defines the sarcomere. Its length at rest is approximately 250 Å ($1 Å = 10^{-10}$ m), meaning that there are about 400 sarcomeres in series within 1 mm of muscle fiber length.

Components of Muscle Fiber

Within each muscle fiber, threadlike fibrils (also called myofibrils, from the Greek *mys,* "muscle"), each wrapped in a membrane (endomysium), are arranged by the hundreds or thousands in parallel. Each of these, in turn, consists of bundles of myofilaments, protein rods that lie parallel to each other. There are two types of such filaments: myosin and actin. Both are elongated polymerized protein molecules. They have the ability to slide along each other, which is the source of muscular contraction (see below). "Stacks" of alternating myosin and actin rods give the appearance of the stripes and bands crossing the muscle fiber, with the myosin filaments bridging the gaps between the ends of adjacent actin filaments, as shown in Figure 3-2.

Table 3-1.   Approximate dimensions of muscle components.

Muscle Components	Diameter	Length
Fiber	$5 * 10^5$ to 10^6 Å	up to 50 cm
Myofibril contains 400–2500 filaments	10^4 to $5 * 10^4$ Å	
Myofilaments:		
Actin	50 to 70 Å	10^4 Å
Myosin	100 to 150 Å	$2 * 10^4$ Å

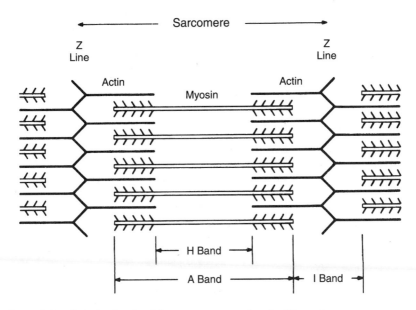

Figure 3-2. Striating bands of the sarcomere crossing the filaments.

Each myofibril contains between 100 and 2,500 myosin filaments, lying side by side, and about twice this many actin filaments. Small projections from the myosin filaments (looking like miniature golf clubs), called crossbridges, protrude toward neighboring actin filaments. The actin filaments are twisted double-stranded protein molecules, wrapped in a double helix around the myosin molecules. In cross section, each myosin rod is surrounded by six actin rods in a regular hexagonal array. This is the contracting microstructure, also occasionally called the elastic element, of the muscle.

Between the myofibrils lie a large number of mitochondria, elongated cells that are the sites of energy production through ATP-ADP metabolism, discussed in Chapter 8. Spaces between the myofibrils are filled with a network of channels, sacs, and cisterns that is connected to the larger tubular system in the z-disks. This is the sarcoplasmic reticulum, the "plumbing and control" system of the muscle. It provides the fluid transport between the cells inside and outside the muscle and also carries chemical and electrical messages. It contains the sarcoplasmic matrix, a fluid that embeds the filaments and contains glucose, glycogen, fat, protein, phosphate compounds, and so forth.

The stripes running across the fiber are of varying widths. As Figure 3-2 shows, the space between adjacent ends of actin proteins is called the H-band. The length of the myosin rods determines the (anisotropic) A-band. The distance between adjacent ends of the myosin rods is the (isotropic) I-band. The middle of the I-band is transversed by the z-disk, which penetrates the actin myofilaments in their centers.

MUSCLE CONTRACTION

The only *active* action a muscle can take is to contract; elongation is brought
about by external forces that stretch the (passive) muscle.

By convention, we distinguish between the resting length of the muscle, its
elongated (stretched) length, and its contracted (shortened) length. These three
conditions are shown in Figure 3-3. In each case, the width of the A-band remains
constant, while the H- and I-bands get narrower in contraction but wider in stretch
compared to the resting state. The sarcomere can be lengthened to about 160% of
its rest length by stretch or shortened to about 60% in contraction.

When the muscle is relaxed, attractive forces between the actin and myosin
filaments are neutralized. After an action potential travels over the muscle fiber
membrane, large quantities of calcium ions are released into the sarcoplasm;
these ions initiate filament activity. According to the "sliding filament" theory
(which is currently well accepted but does not explain all aspects of muscle contraction, especially those in fast shortening), contraction is brought about by the

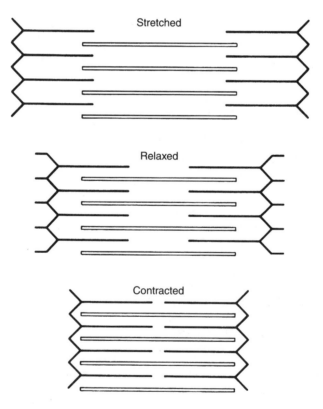

Figure 3-3.   Appearance of a sarcomere within stretched, relaxed, and contracted
muscle.

opposing heads of actin rods moving toward each other. This pulls the z-disks toward the tails of the myosin filaments: the sarcomere shortens. This sliding of filaments is caused by mechanical, chemical, and electrostatic events. Mechanically, forceful interaction takes place at the crossbridges between myosin and actin filaments. The crossbridges are established and released as the actin "ratchets" along the myosin.

Excitation of a muscle follows this sequence.

> **Step 1.** An excitation signal travels along the efferent nervous pathways toward the muscle. The axon of the signal-carrying motor nerve terminates at a motor end plate located in a z-disk, which is the myoneural junction for the motor unit (see below).
>
> **Step 2.** The excitation signal stimulates a depolarizing action of the muscle cell membrane. This allows spread of the action potential (observable in the electromyogram, or EMG) along the sarcoplasmic reticulum.
>
> **Step 3.** The potential triggers the release of calcium into the sarcoplasmic matrix surrounding the filaments of the motor unit.
>
> **Step 4.** This removes the hindrance (tropomyosin) for interactions between actin and myosin filaments (the contracting microstructures of the muscle) through chemical, mechanical, and electrostatic actions.
>
> **Step 5.** Opposing heads of actin rods move toward each other, sliding along the myosin filaments: their heads may meet, bunch up, even overlap. This reduces the length of the sarcomere; the shortening is called *contraction*. Usually, many fibrils shorten at the same time, resulting in a shortening of the whole muscle. (It should be noted here that, properly used, the term contraction refers only to shortening and does not necessarily indicate the generation of tension or force.)
>
> **Step 6.** Rebounding of calcium ions in the sarcoplasmic reticulum switches the contraction activity off, allowing the filaments to relax.

After a contraction, the muscle returns to its resting length, primarily through a recoiling of its shortened filaments, fibrils, fibers, and other connective tissues.

Stretching beyond rest length can be done by forces external to the muscle: either by gravity, by other forces acting from outside the body, or by the action of antagonistic muscle.

Relation Between Muscle Length and Tension

Under a no-load condition, when no external force applies and no internal contraction occurs, the muscle is relaxed at its resting length. Stimulation, with no external load, causes the muscle to actively contract as much as possible until the actin proteins are completely curled around their myosin rods and the heads of opposing actins bunch up against each other. All crossbridges are fully established. This is the shortest possible length of the sarcomere, at about 60% of resting length.

If there is resistance to this *concentric* action, the muscle counteracts it by developing tension between origin and insertion: at the micro level, this means tension between the z-disks. As the sarcomere shortens, the actin rods wrap more around their myosin rods; the curling explains why the actively developed contractile tension weakens as the sarcomere shortens from its resting length. This is sketched in Figure 3-4.

In *eccentric* action, the muscle is elongated by an overpowering external force. The stretch increases its length between insertion and origin, that is, increases the length of sarcomeres. The actin and myosin fibrils are slid along each other by the external stretch force. When the muscle is stretched to about 120% of its resting length, the crossbridges between the actin and myosin rods are in an optimal position to generate contact for active contractile resistance to eccentric stretch. As the fibrils are elongated further, the crossbridge overlap between the protein rods is reduced. At about 160% of resting length, so little overlap remains that no active resistance can be developed internally.

Accordingly, the curve of *active contractile tension* developed within a muscle is minimal at approximately 60% resting length, nearly maximal at resting length, highest at about 120% to 130% of resting length, and then it falls back to minimum at about 160% resting length. (These values apply to an isometric twitch contraction, discussed below under "Strength Depends on Many Factors.")

Yet, the muscle tissues also resist stretch by *passive tension* like a rubber band. During such eccentric action, the muscle tension increases strongly from resting length to a maximum just before the point of muscle or tendon (attachment) breakage, which may happen at about double the resting length.

Consequently, below resting length, the tension within a muscle depends primarily on the active curling of actin around myosin with the use of crossbridges. However, above resting length, the tension in the muscle is the sum of the active and passive strains. This is shown in Figure 3-4. The summation effect explains why we bring "the hand behind the back" when we intend to throw a ball or stone with a lot of force: moving the hand back stretches ("preloads") arm muscles (especially those spanning the shoulder), which facilitates a strong initial acceleration force for the throw.

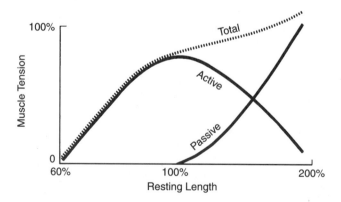

Figure 3-4. Active, passive, and total tension within a muscle at different lengths.

In engineering terms, it is said that muscles exhibit viscoelastic qualities. They are viscous in that their behavior depends both on the amount by which they are deformed and on the rate of deformation. They are elastic in that, after deformation, they return to their original length and shape. However, a muscle is not a purely viscoelastic material because it is also nonhomogeneous, anisotropic, and discontinuous in its mass. Nevertheless, nonlinear elastic theory and viscoelastic descriptors can be used to model major features of muscular performance (Enoka, 1988; Schneck, 1990, 1992).

Viscoelastic theory helps to explain why the tension that can be developed isometrically (statically) by stretching is the highest possible while, in active shortening (dynamic concentric movement), muscle tension is decidedly lower. The higher the velocity of muscle contraction, the faster actin and myosin filaments slide by each other and the less time is available for crossbridges to develop and hold. This reduction in tension capability of the muscle holds true for both concentric and eccentric activities. In eccentric activities, however, where the muscle becomes increasingly lengthened beyond resting length, the total tension resisting the stretch increases with greater length as a result of the summing of active and passive tensions within the muscle.

THE MOTOR UNIT

Muscles are permeated by the axon endings of nerve fibers, often hundreds or even thousands of them; these are the final branchings of the efferent (motor) part of the peripheral nervous system (discussed in Chap. 4). The contact area between the end point of the axon of one motor neuron and the sarcolemma of the muscle is called the motor end plate. Each nerve fiber innervates several muscle fibers, usually hundreds to thousands of fibers through as many motor end plates. These fibers under common control are called a motor unit: they are all stimulated by the same signal. However, the muscle fibers of one motor unit usually do not lie side by side but are in bundles of only a few fibers each, spread throughout the muscle. Thus, "firing" one motor unit does not cause a strong contraction at one specific location in a muscle but rather a weak contraction throughout the muscle (Basmajian and DeLuca, 1985; Buchthal, Guld, and Rosenfalck, 1957; Van Harreveld 1947, 1948).

Considering only one motor unit, we can distinguish between the following types of muscular activity.

Twitch is the single contraction resulting from a single instantaneous stimulus followed by complete relaxation. It lasts about 75 to 220 ms, depending on the muscle. A single twitch consists of a latent period, a period of shortening, a period of relaxation and, finally, a period of recovery. The latent period, typically lasting about 10 ms, shows no reaction yet of the muscle fiber to the motor neuron stimulus. Shortening takes place usually within 40 ms for a fast-twitch fiber. At the end of this period, the muscle element has reached its shortest length and developed tension. (The energy for this process, mostly generated anaerobically, is freed from the ATP complex; see Chap. 8). The heat energy released causes the crossbridges between actin and myosin to undergo a thermal vibration that results in a kind of "ratchet" action, causing the heads of the actin rods to slide toward each other along the myosin filaments. During the relaxation period, commonly about 40 ms for a fast-twitch fiber, the bridges stop oscillating, the bonds between myosin and actin are broken, and the muscle is pulled back to its original length, either by the action of antagonistic muscle or by an external load. (During this period, ATP is resynthesized by ADP. Thus, the primary energy source for muscular contraction is being resupplied.) During the recovery period, again about 40 ms, the metabolism of the muscle is aerobic, with glucose and stored glycogen directly oxidized for the final regeneration of ATP and phosphocreatine. However, if the energy demands on the muscle (through repeated and strong efforts) are beyond the supply capabilities, lactic acid remains as a by-product of anaerobic glycolysis. If so much lactic acid is built up that the breakdown of ATP becomes blocked, the muscle quickly loses its ability to function and must then rest long enough to deplete both the built-up lactic acid and the oxygen debt incurred. Another supplemental explanation is the accumulation of potassium (Kahn and Monod, 1989; Schneck, 1985, 1990, 1992).

Summation (also called superposition) occurs when twitches are initiated frequently one after another, so that a fiber contraction is not yet completely released by the time the next stimulation signal arrives. In this case, the new contraction builds on a level higher than that of completely relaxed fiber, and the contraction achieves accordingly higher contractile tension in the muscle. Such a staircase effect takes place when excitation impulses arrive at frequencies of 10 or more per second. When a muscle is stimulated at or above a critical frequency of about 30 to 40 stimuli per second, successive contractions fuse, resulting in a maintained contraction called tetanus.

Thus, for a single motor unit, the frequency of contractions is controlled by the so-called *rate coding* of the exciting nervous signals. It is of interest to note that, with increasing frequency of contractions building on each other, the force of contraction also increases. In superposition of twitches, the tension generated may be double or triple that of a single twitch and, in full tetanization, may build up to five times the single twitch tension.

MUSCLE FATIGUE

Sufficient supply of the muscle with arterial blood and its unimpeded flow through the capillary bed (see Chap. 7) into the venules and veins are crucial to the muscle because the blood flow determines the ability of the contractile and metabolic processes to continue. Blood brings the needed energy carriers and oxygen, and it removes metabolic by-products, particularly lactic acid and potassium, as well as heat, carbon dioxide, and water liberated during metabolism.

An artery enters the muscle usually at about its half-length and branches profusely. The smallest arteries and their terminal arterioles branch off transversely to the long axes of the muscle fibers, while other arterioles run parallel to individual muscle fibers. Many transverse linkages are present, forming a complex network of blood vessels permeating the muscle tissue. This capillary network is particularly well developed at the motor end plates. The abundance of capillaries in the muscle provides good facilities for the supply of oxygen and nutrition to the muscle cells "bathing" in the interstitial fluid (see Chap. 7) and for the removal of metabolic by-products (see Chap. 8).

The fine blood vessels permeating the muscle are easily compressed by pressure applied to them. A strongly contracting muscle generates strong pressure inside itself, as can be felt by touching a tightened biceps or calf muscle. By this pressure, the muscle compresses its own blood vessels, thus impeding, even shutting off, its own circulation in spite of a reflex increase in systolic blood pressure. Complete interruption of blood flow through a muscle leads to complete muscle fatigue in a few seconds, forcing relaxation.

Insufficient blood flow brings about an accumulation of potassium ions in the extracellular fluid, together with a depletion of extracellular sodium. Combined with intracellular collection of phospate (from the degradation of ATP; see Chap. 8), these biochemical events perturb the coupling between nervous excitation and muscle contraction (see Chap. 4). Depletion of the energy carriers ATP and creatine phospate, along with accumulation of lactate, also occur but are not the main reasons for fatigue.

Muscle fatigue, which occurs more slowly when the muscle is not maximally active, is painfully experienced when one works overhead with raised arms, for instance, fastening a screw in a ceiling. Muscle fatigue in the shoulder muscles makes it impossible to keep one's arms raised after a short time, even though nerve impulses still arrive at the neuromuscular junctions, and the resulting action potentials continue to spread over the muscle fibers.

The higher the requirements are for (isometric) strength exertion of a given muscle, the shorter the period during which this strength can be maintained. Figure 3-5 shows this relation between strength exertion and endurance schematically: a maximal exertion can be maintained for just a few seconds; 50% of strength is available for about 1 min; but less than 20% can be applied for long endurance periods.

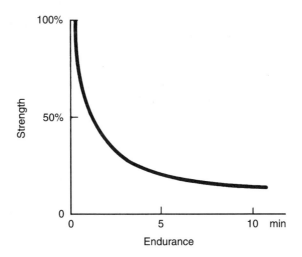

Figure 3-5. Muscle exertion and endurance.

Muscular fatigue can be completely overcome by rest; thus, physiological "fatigue" may be defined operationally (in contrast to psychological "boredom") as a state of reduced physical ability that can be restored by rest.

MUSCULAR ACTIVITIES

In contrast to the foregoing discussion of a single motor unit, the following text relates to the activities of a whole muscle, comprising many motor units. Its contraction activity is controlled by *recruitment coding,* that is, by how many and which motor units are activated at any given instant. Each single motor unit is triggered to contract, and the cooperative effort of the participating motor units determines the contraction of the whole muscle.

Control of the tension of a muscle depends on the number of muscle fibers innervated by one nerve axon: the larger the ratio, the finer the muscle control. For example, in eye musculature, one nerve controls only seven muscle fibers, for an innervation ratio of 1:7, while the quadriceps femoris extending the knee has a ratio of approximately 1:1,000. Gradation of contraction is also controlled by the portion of total muscle fibers excited simultaneously. In general, one cannot voluntarily contract more than two-thirds of all the fibers of a muscle at once; this serves as a safeguard against overstraining the tendon and the attachments at origin and insertion. But contraction of all the fibers at the same time can occur as a result of a proprioceptor reflex; this may strain the muscle or tendon to its total structural tensile capacity and may even tear it.

The action potential spreads along the muscle at speeds of approximately 1 to 5 ms^{-1}. Muscle fibers may consist of a slow-twitch type (also called Type I or red fiber), which has a relatively slow contracting time (80–100 ms until peak is reached) and is well suited for prolonged work because it can work anaerobically. Also, it is triggered by relatively low-rate signals, and is, therefore, often called a low-threshold fiber. In contrast, fast-twitch fibers (also called Type II or white or high-threshold fibers) have relatively short contraction times (about 40 ms) but fatigue rapidly. (Type II-fibers are subdivided again into groups a and b.) The proportion of fiber types in a given individual seems to be genetically determined, but the total number of fibers may be changeable by training (Astrand and Rodahl, 1977); perhaps fibers of one type may be modified into another by use (Astrand and Rodahl, 1986). Each motor unit appears to contain only one type of fiber. Slow fibers are mostly recruited for finely controlled actions, while strong efforts generally involve fast fibers. Both fiber types develop about the same tension. A note of caution: Physiological, biomechanical, and morphological properties of muscle fibers vary over a wide continuum of characteristics. Only as one compares far apart segments with each other does one see distinctions. Furthermore, most research on muscle fiber types has been performed on cats, guinea pigs, and rabbits. In these animals, the muscle fiber distinctions are more clearly apparent than in humans (Basmajian and DeLuca 1985).

The "All-or-None" Principle

All muscle fibers innervated by one motor nerve (i.e., one motor unit) are either fully relaxed or fully contracted. While this is obviously true for the fully excited or fully relaxed states, it does not apply when a muscle is in the initial phase of twitch buildup or during its return to the resting state after attaining contraction. Furthermore, a muscle may be under tension because of external stretch. Hence, the "all-or-none" principle should not be taken to describe truly all conditions of the muscle.

MUSCLE TENSION AND BODY STRENGTH

Unfortunately, the term *strength* is often used in confusing ways: it may refer to the tension *within* a muscle or to the *internal transmission* effects of the muscle tension on links or limbs or to the *external* exertion of force or torque by a body segment to an outside object.

Within the muscle, tension is developed by filament contraction in the longitudinal direction of the muscle fiber, as discussed earlier. The filament tensions combine to a resultant tension of the muscle. The magnitude of the tension depends mostly on the number of muscle fibers involved, that is, on the cross-sectional thickness of the muscle. Maximal tensions reported on human skeletal muscle fall within the range of 16 to 61 N/cm^2. Apparently, training, muscle fiber type, or gender make neither gross nor consistent differences. Accordingly, Enoka (1988) uses 30 N/cm^2 as a typical value and calls it "specific [human muscle] tension." If the muscle cross-section area is known (such as from cadaver measurements or from MRI scans), one can calculate a resultant muscle force. Yet, it is worth remembering that this calculation relies on assumptions about cross section and specific tension values and that it is currently impossible to measure the muscle force directly in the living human.

Muscle tension exists within the muscle-tendon unit, which attaches at origin and insertion to bone links that meet in a common joint. According to this biomechanical model, the tension exerts a torque about the articulation. As discussed under "Internal Transmission," the magnitude of the torque depends on the lever arm and pull angle of the tension vector (see also Chap. 5, "Biomechanics").

The final output of the biomechanical system is the force (or torque) available at the hand, foot, or other body segment for exertion to a resisting object. This object is usually outside the body, as discussed under "External Application of Force and Torque," but it may be an antagonistic muscle. The body (segment) strength available for application to an outside object is of primary importance to the engineer, designer, and manager.

To avoid confusion, the following terms are defined as follows.

Muscle strength *is the maximal tension or force that muscle can develop voluntarily between its origin and insertion.* The best word to refer to this is muscle *tension* (in N/mm^2 or N/cm^2), but the term *strength* (in N) is commonly used.

Internal transmission *is the manner in which muscle tension is transferred inside the body along links and across joints as torque to the point of application to a resisting object.* If several link-joint systems in series constitute the internal path of torque (in Nm or Ncm) transmission, each transfers the arriving torque by the existent ratio of lever arms (*m* and *h* in Figure 3-6) until resistance is met, usually at the point at which the body interfaces with an outside object. This transfer of torques is more complicated under dynamic conditions than in the static case

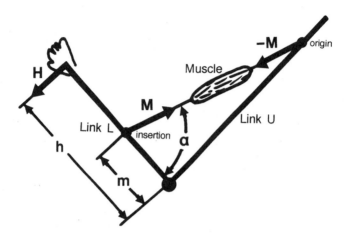

Figure 3-6. The muscle-tendon unit exerting pull forces *M* to links *U* and *L* at origin and insertion.

because of changes in muscle functions with motion and because of the effects of accelerations and decelerations on masses.

Body segment strength *is the force or torque that can be applied by a body segment to an object external to the body.* The segment is usually identified such as hand, elbow, shoulder, back, and so forth. (Strength is in N and torque in Ncm or Nm.)

These definitions are based on the recognition of several facts:

- Measurement of the tension (or force) inside the muscle (or its tendon) is not practicable, at present.
- Therefore, human strength is assessed as the resultant output measured at an instrument externally attached to the body.
- The output measured at the instrument results from the torque that is the product of muscle force pulling on a lever arm (bone link) around an articulation that is spanned by the muscle.
- The tension or force inside the muscle can be calculated from the externally measured output if vector directions, pull angles, and lever arms of the internal transmission system are known. For example, tension within the triceps muscle can be deduced from the force that can be exerted in the hand at right angle to the forearm, as explained in Chapter 5.

Internal Muscle Strength Transmission

The tension developed by a muscular contraction (whether eccentric, isometric, or concentric) tries to pull together two points: the *origin* of the muscle (that attachment closest to the center of the body) and the *insertion* at the opposite end of the muscle-tendon unit (away from the center of the body). This is shown schematically in Figure 3-6.

The origin (or origins, if the muscle has two "heads," such as the biceps) is commonly at the surface of a bone. From there, a short tendon may extend to the muscle belly, where the contractile elements are located. Distal to the muscle, a tendon commonly extends outward. This may be quite long; for example, the tendons that reach from the muscles in the forearm to the digits of the hand can be 20 cm in length. The tendon insertion usually attaches to bone but may also end in strong connective tissues, such as in some of the fingers.

The long bones (representing the links between articulations) are the lever arms at which muscle pulls to generate torque about a body joint. The torque developed depends not only on the strength of the muscle but also on the effective lever arm or on the pull angle. Figure 3-6 depicts these conditions as a simplified scheme of the lower arm link L attached to the upper arm link U, articulated in the elbow joint. The muscle has its origin at the proximate end of U. It is attached at the distance m to the arm link L. It generates a muscle force M, pulling at the angle a. The torque T generated by M depends on the lever arm m and the pull angle a, according to

$$T = m * M * \sin a \tag{3-1}$$

This torque T then counteracts the external force H, assumed to act perpendicularly at its lever arm h, according to

$$T = m * M * \sin a = h * H \tag{3-2}$$

Since the variables m, a, h, and H are measurable, we can solve for the muscle force M.

$$M = \frac{h * H}{m * \sin a} \tag{3-3}$$

(Check Chap. 5 for more detail and further use of biomechanical techniques.)

The simple model depicted in Figure 3-6 shows that the amount of force (H) available at the body interface with an external object depends on internal muscle force (M), lever arms (m and h), and the pull angle (a) which, in turn, depends on the angle between the two links.

External Application of Force and Torque

The original muscle pull (after being internally transmitted to the appropriate body member and transformed in magnitude and direction during that transmission) finally results in a force or torque that can be applied to an outside object: often by hand to a hand tool or to a handle on a box as in load lifting, or by shoulder or back in pushing or carrying, or by the feet in operating pedals or in walking or running.

The quality and quantity of the force or torque transmitted to an outside object depends on many biomechanical and physical conditions, including:

- Body segment employed, for example, hand or foot
- Type of body object attachment, for example, by a simple touch or a surrounding grasp
- Coupling type, for example, by friction or interlocking
- Direction of force/torque vector
- Static or dynamic exertion

Consideration and proper selection of these conditions is a major task of the designer and ergonomist. This is facilitated by clearly distinguishing what happens:

- At the muscle: muscle tension is developed;
- during internal transmission: torques about body joints are transferred across joints
- for external application of segment force/torque to an object.

STRENGTH DEPENDS ON MANY FACTORS

Biomechanical and Central Nervous System (CNS) control factors aside, how much strength one is able to exert depends primarily on which muscle is called into action: How many muscle fibers are involved and how thick are they? At what length is the muscle at the critical demand point?

The amount of tension a muscle can develop is directly related to its size, that is, to its cross-sectional area. The true determiner is the number of muscle fibers and their individual cross-sectional thickness. Strength training increases the thickness of fibers, not their number. Endurance training also increases capillary density and mitochondrial volume (Basmajian and DeLuca, 1985; Enoka, 1988). Training also improves the coordination of motor unit activation directed by the CNS: ". . . muscles are the slaves of their motor neurons" (Basmajian and DeLuca, 1985, p. 431).

Muscle tension also depends on muscle length. Figure 3-4 shows that the maximal *active* effort that a muscle can develop is near its resting length. As the muscle shortens by contraction, less and less additional tension can be achieved until the muscle is totally contracted to approximately 60% of resting length. Alternately, when the muscle is lengthened beyond resting length, the *active* contraction tension is reduced, but *passive* resistance of its elastic tissues against stretch increases. Adding this passive resisting tension to the actively developed tension results in a *total* muscle tension that is largest near 160% of resting length of the muscle.

Another major factor in the amount of strength exerted is the angle at which the muscle pulls on its bony lever arm. Biomechanically, this may be described in terms of leverage, or as the angle between muscle vector direction and lever arm, effective torque, or link position. In everyday language, we call it body posture or limb position. By experience, we learn the skill to position our body segments for achieving that body strength needed to lift a heavy load, to push a big object, or to squeeze the handles of a hand tool.

Movement is also a factor that interacts with muscle mass, muscle length, and vector direction in determining muscle effect.

In one set of cases, we exert our maximal strength with the muscle remaining at a given length during the exertion. In physiology, this is called *isometric* (from the Greek terms *iso,* meaning "the same"; and *metron,* measure referring to length). In physics, this is called *static,* because there is no (change in) movement. In the past, nearly all muscle strength assessments were done under isometric (static) conditions, but there are efforts under way to make dynamic measurements (as discussed below).

In most cases of strength exertion, muscle length changes. Some concentric motions are quite fast, as in running, throwing a ball, shoveling, or hammering. In such quick muscle contractions, the dynamic muscle output may be rather different from statically applied muscle strength. In contrast, eccentric muscle efforts are usually rather slow and, therefore, isometric measurements should provide appropriate information about eccentric muscle strength.

REGULATION OF STRENGTH EXERTION

One way to understand the factors involved in the exertion of voluntary strength is to use the model in Figure 3-7. The control initiatives generated in the central nervous system start with calling up an "executive program" (to use a computer analogy) which, whether innate or learned, exists for all normal muscular activities, such as walking or pushing and pulling objects. The general program is modified by "subroutines" appropriate for the specific case, such as walking quickly upstairs or pulling hard or pushing carefully. (The general or specific skill programs are occasionally called *engrams.*) The outputs from these routine programs, in turn, are modified by motivation, which determines how (and how much of the structurally possible) strength will be exerted under the given conditions. A listing of circumstances that may increase or decrease one's willingness to exert is given in Table 3-2.

The result of these complex interactions manifests itself in the excitation signals E transmitted along the efferent nervous pathways to the motor units involved, where they trigger muscle contractions. The actual tension developed in the muscle depends on the rate and frequency of signals received, on muscle size and motor units involved and, possibly, on existing fatigue in the muscle left from previous contractions, as discussed earlier.

The output of the muscular contraction is modified by the existing biomechanical conditions, especially lever arms and pull angles, as discussed. These conditions would change in the course of dynamic activities but are assumed constant in a static effort.

Thus, the output of this complicated chain of controllers, feedforward signals, controlled elements, and modifying conditions is the body segment strength measured at the interface between the body segment involved and the measuring device (the object against which strength is exerted). Of course, the assumption is that, at any moment, at least as much resistance is available at this interface as can be exerted by the person. If this is not the case, that is, if Newton's third law is violated, no reliable assessment of body strength can be performed.

The model also shows a number of feedback loops through which the muscular exertion is monitored for control and modification. The first feedback loop, $F1$, is in fact a reflexlike arc that originates at proprioceptors, such as Ruffini organs in the joints (signaling location), Golgi tendon-end organs (indicating

AFFERENT FEEDBACK LOOPS   **MEASUREMENT OPPORTUNITIES**

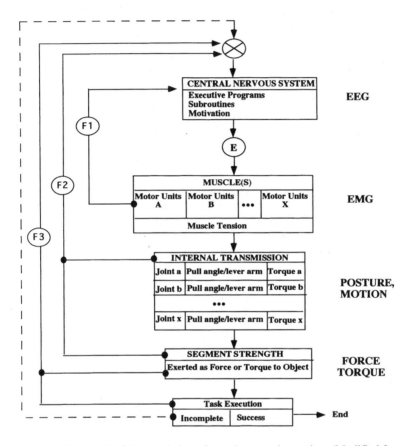

Figure 3-7. Model of the regulation of muscle strength exertion: (Modified from Kroemer, 1979.) *E*: efferent excitation impulses generated in the CNS; *F*: afferent feedback loops.

changes in muscle tension), and muscle spindles (indicating length). These interoceptors and their signals are not under voluntary control and influence the signal generator in the spinal cord very quickly (but how they do it is not well understood).

The other two feedback loops originate at exteroceptors and are routed through a comparator modifying the input signal into the central nervous system. The second loop, *F2*, originates at receptors signaling events related to touch, pressure, and body position in general (kinesthetic signals), such as one feels when pulling on a handle: body position is monitored, together with the sensations of pressure in the hand coupled with the instrument and of pressure felt at

Table 3-2. Factors likely to increase (+) or decrease (–) maximal muscular performance.

	Likely Effect
Feedback of results to subject	+
Instructions on how to exert strength	+
Arousal of ego involvement, aspiration	+
Pharmaceutical agents (drugs)	+
Startling noise, subject's outcry	+
Hypnosis	+
Setting of goals, incentives	+ or –
Competition, contest	+ or –
Verbal encouragement	+ or –
Fear of injuries	–
Spectators	?
Deception	?

the feet through which the chain of force vectors is transmitted in a standing position. The third feedback loop, $F3$, originates at exteroceptors; it signals such events as sounds and vision related to the effort to the comparator; for example, the sounds or movements generated in the experimental equipment by the exertion of strength, the pointer of an instrument that indicates the strength applied, or the experimenter or coach giving feedback and exhortation to the subject, depending on the status of the effort.

MEASUREMENT OPPORTUNITIES

For the sake of clarity, let us continue to separate the model into three sections: feedforward, feedback, and output. The *feedforward* part includes the contracting muscle and the internal transmission of muscle strength (including the nervous control); this is followed by the sensory *feedback* to the comparator in the brain. The *output* is the body strength exertion at the interface with an external object (work tool or measuring instrument). Each section provides opportunities to measure.

Measuring During Feedforward

Considering the feedforward section of the model, it becomes apparent that, currently, we still lack suitable means to measure the executive programs, the subroutines, or the effects of will or motivation on the signals (E) generated in the central nervous system. Only very general information can be gleaned from electroencephalograms (EEGs).

Better observable events are the efferent excitation impulses that travel along the motor nerves to the muscles, where they spread from the motor end plates. The electric events associated with the contraction of ("Muscle" in Fig. 3-7) motor units can be observed through (intrusive or surface) electrodes and recorded and evaluated via electromyograms. Some experimental techniques are available to measure directly the tensions within muscle filaments, fibrils, fibers, or groups of muscles in situ.

Electromyography

One technique for assessing the events at the muscle uses the electromyogram (EMG): the electrical signals associated with the contraction process are recorded via surface or indwelling electrodes. The intent is to relate exerted muscle strength to the magnitude of the EMG signal. For this, it is necessary that the EMG amplitudes be calibrated against known contraction levels. This calibration still poses a major challenge although EMG measurements have been taken since the 1930s.

Interpretation of an electromyogram, the record of the electrical signals associated with the activation of motor units in muscle in terms of frequency and amplitude, relies on several basic assumptions. One assumption is that the signals stem from the same motor units. A surface electrode can be used for isometric contractions because, in this static case, the muscle does not move under the electrode on the skin; in dynamic muscle use, indwelling (wire or needle) electrodes can follow the moving muscle. Another assumption concerns the relationship between EMG amplitude and muscle effort. In the isometric case, calibration often shows a nearly linear increase in EMG intensity from rest to maximal voluntary muscle exertion; but, in dynamic muscle use, the EMG-effort relationship is complex and difficult to establish. Another complication comes with exertion time, when rate and recruitment coding of motor unit excitation often vary; this is particularly the case with muscle fatigue, where the EMG generally shows a shift toward lower frequencies, usually together with an increase in amplitude (Basmajian and DeLuca, 1985; Chaffin and Andersson, 1991; Soderberg, 1992). Thus, electromyography is not an "easy" technique to assess muscle strength.

Measuring at the Muscle-Tendon Unit

There are several experimental approaches to recording muscle strength directly. One uses a thin optic fiber pushed through a muscle. The light transmission is affected by deformation of the fiber by muscle pressure, which itself increases with stronger muscle tension. This measurement so far has been done in the leg muscle of a hare (Komi et al., 1996).

At the occasion of surgery to relieve carpal tunnel pressure, a finger flexor tendon can be pulled clear from its surrounding tissues and threaded through a calibrated measurement device (formed like a belt buckle). The patient is asked to contract to associated muscle, and the force in the tendon can be measured (Schuind et al., 1992). A similar transducer has also been attached to the Achilles tendon of a human foot to measure tendon loading during jumps (Fukashiro et al., 1995).

Measuring Mechanical Advantages

The physical conditions during internal transmission can be observed and recorded in terms of body segment angles and lever arms. However, the mechanical advantages internal to the body are much more difficult and often practically impossible to record and control, especially during motions as opposed to a static posture. (See the discussion of statics and dynamics below.) This concerns, for example, lever arms of the tendon attachments and pull angles within the muscle-tendon link system.

These examples indicate current problems and future possibilities in the measurement of human voluntary strength. Apparently, the result of each signal, process, or condition in the feedforward loop of the model may be modified by following factors, processes, or conditions that are largely unknown or difficult to interpret. For example: The components of the "decision making" in the CNS are not discernible from the EEG; the consequences of contraction signals observable in the EMG depend on the conditions of the muscle; the effects of muscular contractions are modified by physical conditions of the internal transmission, such as mechanical advantages. (Yet, if it becomes feasible to measure the internal force directly, our earlier definition of muscle strength could be abolished in favor of "the maximal voluntary force that a muscle can exert along its length.")

Measuring Feedback

The feedback loops in the nervous system offer some interesting possibilities. Yet the afferent pathways from interoceptors are anatomically and functionally associated with the feedforward paths for the efferent impulses. Hence, it is practically impossible (with current technology) to distinguish the electric events associated with the feedback signals from those associated with the feedforward signals. Measurement and control of the first two feedback loops ($F1$ and $F2$) is not easily accomplished; currently, in fact, they are not practicable for strength measurements.

The third feedback loop ($F3$) starting at exteroceptors, such as the eyes and ears, can be experimentally manipulated, for example, by providing (or holding back) information about test results to the subjects (Kroemer and Marras, 1980). Experimenters, trainers, and coaches use this feedback to exhort enhanced performance.

Measuring Output

With current techniques, the least ambiguous and best accessible measurement point is at the interface between the body and the instrument that records the resultant output of all components in the loops: "body (or segment) strength is what is measured externally." (This is reminiscent of the similarly unsatisfying definition of intelligence as "what is measured in an intelligence test.")

The resulting output of this complex system is clearly definable and measurable in terms of amount and direction of torque or force exerted, over time, to the measuring device. The conditions external to the body are fairly easy to describe: location of the interface between body and measuring device; positioning and support of the body, including coupling at hands or feet; temperature; humidity; and so forth.

The simplicity of measuring the resulting output (of this complicated system of generating and controlling and internally transmitting voluntary muscle strength) at an external instrument is one good reason for doing so. The practical usefulness of that simple information for trainer, coach, physician, engineer, and "owner" is another reason.

ASSESSMENT OF BODY SEGMENT STRENGTH: STATICS AND DYNAMICS

As discussed, generation of strength is a complex procedure of myofilament activation through nervous feedforward and feedback control and the use of mechanical leverages within the body. It may involve substantial shortening or lengthening of muscle in a concentric or eccentric effort; or there may be no perceptible change in length—that is, the effort is isometric.

Human body (segment) strength is measured mostly at the interface between the body and the measurement device, not directly at the muscle. For practical applications, the data recorded at the instrument show what is available for work and leisure and sport activities. Mechanically, the main distinction between muscle actions is whether they are static or dynamic.

Static Strength

As already mentioned, an isometric muscle contraction (a physiological description) generates—probably after some initial muscle shortening—a static condition (a physical description). When there is no change in muscle length and segment position during the isometric effort, then involved body segments do not move; all forces acting within the system are in equilibrium, as Newton's first law requires. Therefore, the physiological isometric case is equivalent to the static condition in physics.

The static condition is theoretically simple and experimentally well controllable. It allows rather easy measurement of muscular effort and, therefore, most of the information currently available on human strength describes outcomes of static (isometric) testing. Accordingly, most of the tables on strength in this chapter contain static data.

Besides the simple convenience of dealing with statics, measurement of isometric strength appears to yield (for most cases except fast ballistic-impulse type of contractions, such as throwing) a reasonable estimate of the maximally possible muscle exertion. Exertion-velocity curves (seemingly based on works first published by Best and Taylor in 1973 and by Winter in 1979) indicate that the largest tension or force is indeed developed at zero velocity of muscle shortening or lengthening, that is, in the isometric case; see Figs. 3-8 and 3-9.

"Dynamic Strength"

A few more words on terminology: As just mentioned, the physiologic "isometric" exertion generates a static condition. The term *variometric* or, less descriptively, *anisometric* refers to a change in muscle length that produces motion, that is, a dynamic condition. Within the physics field of dynamics, one talks about

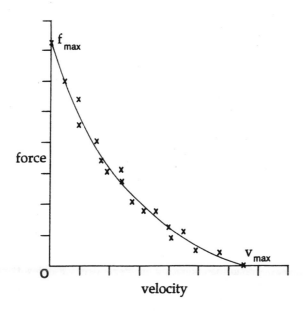

Figure 3-8. Force-velocity relationship of muscle (attributed by Chaffin and Andersson, 1991, to Best and Taylor, 1973).

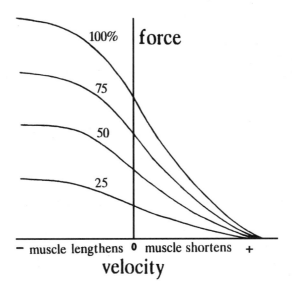

Figure 3-9. Force-velocity relationship of muscle (attributed by Chaffin and Andersson, 1991, to Winter, 1979).

kinematics when investigating motion, and *kinetics* when considering the force that causes motion according to Newton's first law.

Dynamic muscular efforts are more difficult to describe and control than static contractions. In dynamic activities, muscle length changes and, therefore, involved body segments move. This results in displacement. The amount of travel is relatively small at the muscle but is often amplified at the point of exertion to the outside, for example, at the hand or foot.

The time derivatives (velocity, acceleration, and jerk) of displacement are of importance both for the muscular effort and the external effect. For example, change in velocity determines impact and force, as expressed in Newton's second law.

Definition and experimental control of dynamic muscle exertions are much more complex tasks than the experimenter encounters in static testing. Various new schemes for independent and dependent experimental variables could be developed, but it is desirable to include the traditional isometric approach. Such a system has been presented for dynamic and static efforts (Kroemer et al., 1989).

Independent variables are those that are purposely manipulated during the experiment to assess resulting changes in the dependent variables. For example, if we set the displacement (muscle length change) to zero—the isometric condition—we may measure either the magnitude of force generated or the number of repetitions that can be performed until force is reduced because of muscular fatigue. This case is described in Table 3-3. Of course, with no displacement, its time derivatives—velocity, acceleration, and jerk—are also zero.

Alternatively, we may choose to control velocity as an independent variable, that is, the rate at which muscle length changes. If velocity is set to a constant value, we speak of an isokinematic muscle strength measurement. (Note that this isovelocity condition is often mislabeled "isokinetic.") Time derivatives of constant velocity, acceleration and jerk, are zero. Mass properties are usually controlled in isokinematic tests. The variables displacement, force, and repetition may either be chosen as dependent variables or become controlled independent variables. Most likely, force and/or repetition are chosen as the dependent test variables.

Following the scheme laid out in Table 3-3, tests can also be devised in which acceleration or its time derivative, jerk, are kept constant. These test conditions are theoretically possible, but so far they have not been commonly applied.

For some tests, the amount of muscle tension (force) is set to a constant value. In this "isoforce" test, mass properties and displacement (and its time derivatives) are likely to become controlled independent variables, and repetition is likely to become a dependent variable. The isotonic condition is, for practical reasons, often combined with an isometric condition, such as in holding a load motionless.

Note that the term *isotonic* has often been wrongly applied. Some older textbooks have cited examples of lifting and lowering of a constant mass (weight) as typical of isotonics. This is physically false for two reasons. The first follows from Newton's second law: the change from acceleration to deceleration of a mass requires application of changing (not constant) forces. The second fault lies in overlooking the changes that occur in the mechanical conditions (pull angles and lever arms) under which the muscle functions during the activity (See

Table 3-3. Techniques to measure motor performance by selecting specific independent and dependent variables.

Names of Technique / Variables	Isometric (Static)		Isovelocity Isokinematic (Dynamic)		Isoacceleration (Dynamic)		Isojerk (Dynamic)		Isotonic Isoforce (Static or Dynamic)		Isoinertial (Static or Dynamic)		Free Dynamic	
	Indep.	Dep.	Indep.	Dep.	Indep.	Dep.	Indep.	Dep.	Indep.	Dep.	Indep.	Dep.	Indep.	Dep.
Displacement linear/angular	constant* (zero) [boxed]		C	or X	C	or X	C	or X	C	or X	C	or X		X
Velocity, linear/angular	O		constant [boxed]		C	or X	C	or X	C	or X	C	or X		X
Acceleration, linear/angular	O		O		constant [boxed]		C	or X	C	or X	C	or X		X
Jerk, linear/angular	O		O		O		constant [boxed]		C	or X	C	or X		X
Force, Torque	C	or X	C	or X	C	or X	C	or X	constant [boxed]		C	or X		X
Mass, Moment of Inertia	C		C		C		C		C		constant [boxed]		C	or X
Repetition	C	or X	C	or X	C	or X	C	or X	C	or X	C	or X	C	or X

Legend
C = variable can be controlled
* = set to zero
O = variable is not present (zero)
X = can be dependent variable

The [boxed] constant variable provides the descriptive name.

Chap. 5 for more detail.) Hence, even if there were a constant force to be applied to the external object (which is not the case), the changes in mechanical advantages would result in changes in muscle tone. It is certainly misleading to label all dynamic activities of muscles isotonic, as is unfortunately still done occasionally.

In the isoinertial test, the external mass is set to a constant value. In this case, repetition of moving such constant mass (as in lifting) may be either a controlled independent variable or, more likely, a dependent variable. Also, displacement and its derivatives may become dependent outputs. Force (or torque) applied is likely to be a dependent value, according to Newton's second law.

Table 3-3 also contains the most general case of motor performance measurement, labeled *free dynamic*. In this case, the independent variables displacement (and its time derivatives) and force are left to the free choice of the subject. Only mass and repetition are usually controlled but may be used as dependent variables. Force, torque, or some other performance measure is likely to be chosen as a dependent output.

The foregoing discussion indicates that dynamic tests indeed require more effort to describe and control than static (isometric) measurements. This complexity explains why, in the past, dynamic measurements other than isokinematic and isoinertial testing have rarely been performed in the laboratory. On the other hand, if one is free to perform as one pleases, such as in the free dynamic test common in sports, very little experimental control can be executed.

STRENGTH MEASUREMENT DEVICES

Devices to measure force or torque (moment) consist of several components. The first is the sensor, the element that experiences the strain generated by force or torque application, and the second is the converter, the element that changes the strain into a measurable output. Both elements together are often called the *transducer*. A typical transducer consists of a deformable object, such as a metal beam, which is bent (usually almost imperceptibly) under the effort exerted on it. A strain gauge is commonly employed to convert the deformation into an electrical signal that is analog to the strain and deformation. (In this context, it is rather interesting to note that force and torque are not basic units in the international measurement system but are derived from the product of mass and acceleration; see Chap. 5.) There are no instruments available at this time that are directly sensitive to force or torque; all rely on the sensor-converter technique.

The output of the transducer is usually fed through an amplifier so that the signal can be easily transmitted and used. The next element in the measurement device often is an indicator (display). This is typically a pointer displaced from its zero position according to the strength of the signal received from the amplifier; it may move over a stationary scale or leave a permanent mark on a strip chart. The displacement of these indicators is analog to the signal. Other indicators are digital, giving the signal in discrete numbers. Usually, the system includes a data storage device, a recorder or computer, or both, which may be in series or parallel with the indicator. (In some cases, an indicator is not used in the system with a recorder, yet it is useful for observing the data flow so that obvious problems can be detected during the test.) The data analysis is strictly not part of the measurement chain. However, selection of the statistical technique used may have determining effects on the selection of the measurement device itself.

One important aspect is unfortunately too often overlooked: whether home-built or purchased, the measurement device must be calibrated to assure that the same input results in the same known output in each test. (It is discouraging to see in a laboratory a measurement device in use that has not been checked and calibrated for long periods of time.)

THE STRENGTH TEST PROTOCOL

After choosing the type of strength test to be performed, as well as the measurement techniques and the measurement devices, an experimental protocol must be devised. This includes the selection of subjects and their information and protection; the control of the experimental conditions; the use, calibration, and maintenance of the measurement devices; and (usually) the avoidance of training and fatigue effects. Regarding the selection of subjects, care must be taken that the subjects participating in the tests are, in fact, a representative sample of the population for which data are to be gathered. As for management of the experimental conditions, the control over motivational aspects is particularly difficult. It is widely accepted (outside sports and medical function testing) that the experimenter should not give exhortations and encouragements to the subject (Caldwell et al., 1974). The so-called Caldwell regimen was meant for isometric testing but can be adapted for dynamic tests as indicated by an asterisk (*). The following listing contains its main features with some minor amendments.

Static (body segment) strength is defined as the capacity to produce torque or force by a maximal voluntary isometric exertion. (Dynamic body segment strength is then the capacity to produce torque or force by a maximal voluntary variometric exertion.*) Strength has vector qualities and, therefore, should be described by magnitude and direction.

1. Static strength is measured according to the following conditions:
   (a) Static strength is assessed during a steady exertion sustained for 4 sec.
   (b) The transient periods of about 1 sec each, before and after the steady exertion, are disregarded.
   (c) The strength datum is the mean score recorded during the first 3 sec of the steady exertion.
2. Instructions are to be given to the subject.
   (a)* The subject should be informed about the test purpose and procedures.
   (b)* Instructions to the subject should be kept factual and should not include emotional appeals.
   (c) The subject should be instructed to increase to maximal exertion (without jerk) in about 1 sec and maintain this effort during a 4-sec count.
   (d)* Inform the subject during the test session about his/her general performance in qualitative, noncomparative, positive terms. Do not give instantaneous feedback during the exertion.

    (e)* Rewards, goal setting, competition, spectators, fear, noise, and so forth, can affect the subject's motivation and performance and, therefore, should be avoided.
3.* The minimal rest period between related efforts should be 2 min, more if symptoms of fatigue are apparent.
4. Describe the conditions existing during testing:
    (a)* Body parts and muscles chiefly used.
    (b)* Body position.
    (c)* Body support/reaction force available.
    (d)* Coupling of the subject to the measuring device (to describe location of the strength vector).
    (e)* Strength measuring and recording device.
5. Subject description:
    (a)* Population and sample selection.
    (b)* Current health and status: medical examination and questionnaire are recommended.
    (c)* Gender.
    (d)* Age.
    (e)* Anthropometry (at least height and weight).
    (f)* Training related to the strength testing.
6. Data reporting:
    (a)* Mean (median, mode).
    (b)* Standard deviation.
    (c)* Skewness.
    (d)* Minimum and maximum values.
    (e)* Sample size.

Note that, in some isometric (static) tests, the measurements have been confined to taking only the peak strength observed during the effort. The peak is up to one-third larger than the average measure over an exertion period of 3 sec; see item 1(c) above.

DESIGNING FOR BODY STRENGTH

The engineer or designer concerned with human strength has to make a number of decisions. These include:

Is the use mostly static or dynamic? If static, information about isometric strength capabilities, listed below, can be used. If dynamic, additional considerations apply concerning, for example, physical endurance, (circulatory, respiratory, metabolic) capabilities of the operator, and prevailing environmental conditions. Physiologic and ergonomic texts (e.g., by Astrand and Rodahl, 1986; Kroemer, Kroemer, and Kroemer-Elbert, 1994; Winter, 1990) provide such information.

Is the exertion by hand, by foot, or with the whole body? For each, specific design information is available. If a choice is possible, it must be based on physiologic and ergonomic considerations to achieve the safest, least strenuous, and most efficient performance. In comparison to hand movements over the same distance, foot motions consume more energy, are less accurate and slower, but are stronger.

Is a maximal or a minimal strength exertion the critical design factor?

Maximal user output usually determines the structural strength of the object, so that a handle or a pedal may not be broken by the strongest operator. The design value is set, with a safety margin, above the highest expected strength application.

Minimal user output is that effort expected from the weakest operator that still yields the desired result, so that a door handle or brake pedal can be successfully operated or a heavy object be moved.

The range between those maximal and minimal values includes all expected applications. An "average" user exertion is usually of no value for the designer.

Most body segment strength data are available for static (isometric) exertions. They provide reasonable guidance also for slow motions although they are probably a bit too high for concentric motions and too low for eccentric motions. Of the little information available for dynamic strength exertions, much is limited to isokinematic (constant-velocity) cases. As a general rule, strength exerted in motion is less than measured in static positions located on the path of motion.

Measured strength data are often treated statistically as if they were normally distributed and reported in terms of averages (means) and standard deviations. This allows the use of common statistical techniques to determine data points of special interest to the designer, as discussed in detail in Chapter 1. In reality, body segment strength data are often skewed rather than in a bell-shaped distribution. This is not of great concern, however, because the data points of interest are usually the extremes: either the maximal forces or torques that the equipment must be able to bear without breaking or the minimal exertions that even "weak" persons are able to generate. These can be identified as given percentile values at the ends of the distribution: for weak exertions, often the fifth percentile is selected. This can be done either by calculation (see Chap. 1) or by estimation.

The Use of Tables of Exerted Torques and Forces

There are many sources for data on body strengths that operators can apply. While these data indicate orders of magnitude of forces and torques, the exact numbers should be viewed with great caution because they were measured usually on diverse and small subject groups under widely varying circumstances. It is advisable to take body strength measurements on a sample of the intended user population to verify that a new design is operable.

Note that thumb and finger forces, for example, depend decidedly on skill and training of the digits as well as the posture of the hand and wrist. Hand forces (and torques) also depend on wrist positions, and on arm and shoulder posture. Exertions with arm, leg, and body (shoulder, backside) depend much on the posture of the body and on the support provided to the body (i.e., on the reaction force in the sense of Newton's third law) in terms of friction or bracing against solid structures. Figure 3-10 and Table 3-4 illustrate this: both were derived from the same set of empirical data but extrapolated to show the effects of location of the point of force exertion, body posture, and friction at the feet on horizontal push (and pull) forces applied by male soldiers.

	Force-plate[1] height	Distance[2]	Force, N	
			Mean	SD
	50	80	664	177
	50	100	772	216
	50	120	780	165
	70	80	716	162
	70	100	731	233
	70	120	820	138
	90	80	625	147
	90	100	678	195
	90	120	863	141
	Percent of shoulder height		Both hands	
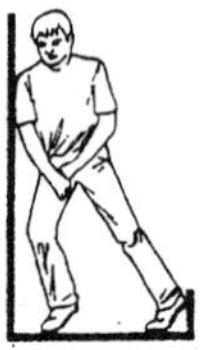	60	70	761	172
	60	80	854	177
	60	90	792	141
	70	60	580	110
	70	70	698	124
	70	80	729	140
	80	60	521	130
	80	70	620	129
	80	80	636	133
	Percent of shoulder height			
	70	70	623	147
	70	80	688	154
	70	90	586	132
	80	70	545	127
	80	80	543	123
	80	90	533	81
	90	70	433	95
	90	80	448	93
	90	90	485	80
	Percent of shoulder height		Both hands	
			Both hands	
Force plate	100 percent of shoulder height	50	581	143
		60	667	160
		70	981	271
		80	1285	398
		90	980	302
		100	646	254
			Preferred hand	
		50	262	67
		60	298	71
		70	360	98
		80	520	142
		90	494	169
		100	427	173
		Percent of thumb-tip reach*		
	100 percent of shoulder height	50	367	136
		60	346	125
		70	519	164
		80	707	190
		90	325	132
		Percent of span**		

[1]Height of the center of the force plate – 20 cm high by 25 cm long – upon which force is applied.
[2]Horizontal distance between the vertical surface of the force plate and the opposing vertical surface (wall or footrest, respectively) against which the subjects brace themselves.

*Thumb-tip reach – distance from backrest to tip of subject's thumb as arm and hand are extended forward.
**Span – the maximal distance between a person's fingertips when arms and hands are extended to each side.

Figure 3-10. Mean horizontal push forces (and standard deviations), in newtons, exerted by standing men with hand, shoulder, and back (adapted from NASA, 1989).

Table 3-4. Horizontal push and pull forces, in newtons, that male soldiers can exert intermittently or for short periods of time (adapted from U.S. Army Missile Command, MIL-STD 1472, 1981b).

Horizontal force*; at least	Applied with**	Condition (μ: coefficient of friction at floor)
100 N push or pull	Both hands or one shoulder or the back	With low traction, $0.2 < \mu < 0.3$
200 N push or pull	Both hands or one shoulder or the back	With medium traction, $\mu \sim 0.6$
250 N push	One hand	If braced against a vertical wall 51–150 cm from and parallel to the push panel
300 N push or pull	Both hands or one shoulder or the back	With high traction, $\mu \sim 0.9$
500 N push or pull	Both hands or one shoulder or the back	If braced against a vertical wall 51–180 cm from and parallel to the panel or If anchoring the feet on a perfectly nonslip ground (like a footrest)
750 N push	The back	If braced against a vertical wall 600–110 cm from and parallel to the push panel or If anchoring the feet on a perfectly nonslip ground (like a footrest)

*May be nearly doubled for two and less than tripled for three operators pushing simultaneously. For the fourth and each additional operator, not more than 75% of their push capability should be added.

**See Figure 3-10 for examples.

It is obvious that the amount of strength available for exertion to an object outside the body depends on the weakest part in the chain of strength-transmitting body parts. Hand pull force, for example, may be limited by a person's finger strength, or shoulder strength, or low-back strength; or it may be limited by the reaction force available to the body, as in Newton's third law. Figure 3-11 helps in determining where the critical body segment is in that sequence of torques about body joints. Starting at the point of external exertion—for example, at the hand—one assesses the strength requirements joint by joint along the arm, shoulder, and back. Often, the lumbar back area is the weak link, as evidenced by the large number of low-back pain cases reported in the literature; see Chapter 10 specifically for lifting and other manual material movement.

Human Strength Evaluation

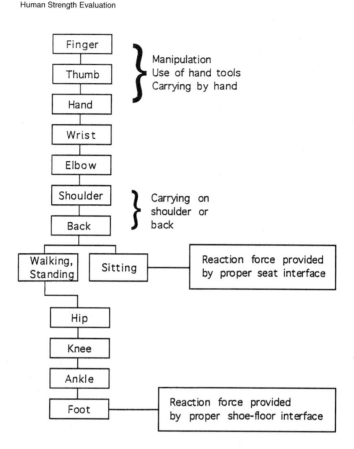

Figure 3-11. Determining the critical body segment strength for manipulating and carrying.

To a sitting person, the reaction countering the forces actively exerted through upper body and arms is provided largely by the seat, although some support may be gathered from the floor via the legs. A walking or standing person receives all support from the ground up, of course, and hip or knee joint strength may limit the ability to make hard efforts, such as lifting a load on the back. A slippery surface may make it impossible to push a heavy object sideways with the shoulder. These examples demonstrate how important it is to provide proper body support at seat or ground.

Designing for Hand Strength

The human hand is able to perform a large variety of activities, ranging from those that require fine control to others that demand large forces. (The feet and legs are capable of more forceful exertions than the hands, as discussed in the following section.)

One may divide hand tasks in this manner:

Fine manipulation of objects, with little displacement and force. Examples are writing by hand, assembly of small parts, and adjustment of controls.

Fast movements to an object, requiring moderate accuracy to reach the target but, once there, fairly small force exertion. An example is the movement to a switch and its operation.

Frequent movements between targets, usually with some accuracy but little force, such as taking parts from bins and assembling them.

Forceful activities with little or moderate displacement, such as turning a hand tool against resistance.

Forceful activities with large displacements, such as hammering.

Accordingly, there are three major types of requirements: for accuracy, for displacement, and for strength exertion. Design for the first two tasks is described in greater detail by Kroemer, Kroemer, and Kroemer-Elbert (1994), in Chapters 8 through 11 of their book *Ergonomics*.

Of the digits of the hand, the thumb is the strongest and the little finger the weakest. Gripping and grasping strengths of the whole hand depend on the coupling used between the hand and the handle; see Figure 3-12. The forearm can develop fairly large twisting torques. Large force and torque vectors can be exerted with the elbow at about right angle, but the strongest pulling/pushing forces toward/away from the shoulder can be exerted with the extended arm, provided that the trunk can be braced against a solid structure. Torque about the elbow depends on the elbow angle, as depicted in Figure 3-13 and, in more detail, in Figure 3-14. (Note that the calculation of percentiles is explained in detail in Chap. 1).

Obviously, forces exerted with the arm and shoulder muscles are largely determined by body posture and body support. Likewise, finger forces depend on the finger joint angles, as listed in Tables 3-5 and 3-6. Table 3-7 provides detailed information about manual force capabilities measured in male students and machinists. Female students developed between 50% and 60% of the digit strength of their male peers but achieved 80% to 90% in pinches.

Designing for Foot Strength

If a person stands at work, fairly little force and only infrequent operations of foot controls should be required because, during these exertions, the operator has to stand on the other leg. For a seated operator, however, operation of foot controls is much easier because the body is largely supported by the seat. Thus, the feet can move more freely and, given suitable conditions, can exert large forces and energies.

A typical example of such an exertion is pedaling a bicycle: all energy is transmitted from the leg muscles through the feet to the pedals. For normal use, the pedals should be located directly underneath the body, so that the body weight above them provides the reactive force to the force transmitted to them. Placing

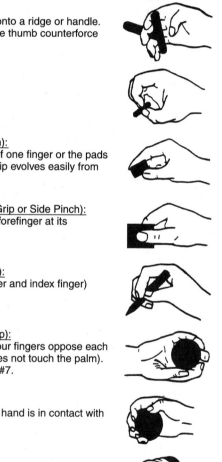

1. Digit Touch:
    One digit touches an object without holding it.

2. Palm Touch:
    Some part of the inner surface of the hand touches the object
    without holding it.

3. Finger Palmar Grip (Hook Grip):
    One finger or several fingers hook(s) onto a ridge or handle.
    This type of finger action is used where thumb counterforce
    is not needed.

4. Thumb-Fingertip Grip (Tip Pinch):
    The thumb tip opposes one fingertip.

5. Thumb-Finger Palmar Grip (Pad Pinch):
    Thumb pad opposes the palmar pad of one finger or the pads
    of several fingers near the tips. This grip evolves easily from
    coupling #4.

6. Thumb-Forefinger Side Grip (Lateral Grip or Side Pinch):
    Thumb opposes the radial side of the forefinger at its
    middle phalanx.

7. Thumb–Two-Finger Grip (Writing Grip):
    Thumb and two fingers (often forefinger and index finger)
    oppose each other at or near the tips.

8. Thumb-Fingertips Enclosure (Disk Grip):
    Thumb pad and the pads of three or four fingers oppose each
    other near the tips (object grasped does not touch the palm).
    This grip evolves easily from coupling #7.

9. Finger-Palm Enclosure (Enclosure):
    Most, or all, of the inner surface of the hand is in contact with
    the object while enclosing it.

10. Grasp (Power Grasp):
    The total inner hand surface is grasping the (often cylindrical)
    handle, which runs parallel to the knuckles and generally
    protrudes from one side or both sides of the hand.

Figure 3-12. Couplings between hand and handle (based on the 1986 taxonomy
described in Coupling the Hand with the Handle: An Improved Notation of Touch, Grip,
and Grasp by K.H.E. Kroemer, *Human Factors* 28:337–339.)

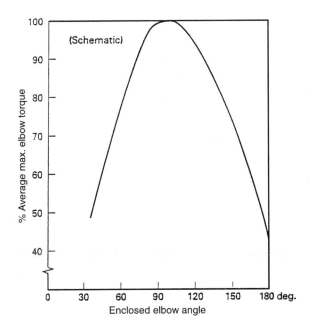

Figure 3-13. Relation of Elbow Angle and Elbow Torque. (With permission of the publisher from Kroemer, K.H.E., Kroemer, H.B., and Kroemer-Elbert, K.E. (1994). *Ergonomics: How to Design for Ease and Efficiency.* Englewood Cliffs, NJ: Prentice-Hall. All rights reserved.)

Table 3-5. Mean forces (and standard deviations), in N, exerted by nine subjects in fore, aft, and down directions with the fingertips, depending on the angle of the proximal interphalangeal (PIP) joint. (Reprinted with permission of the publisher from Kroemer, K. H. E., Kroemer, H. B., and Kroemer-Elbert, K. E. (1994). *Ergonomics: How to Design for Ease and Efficiency.* Englewood Cliffs, NJ: Prentice–Hall. All rights reserved.)

	PIP joint at 30 deg			*PIP joint at 60 deg*		
Direction:	**Fore**	**Aft**	**Down**	**Fore**	**Aft**	**Down**
Digit						
2 Index	5.4 (2.0)	5.5 (2.2)	27.4 (13.0)	5.2 (2.4)	6.8 (2.8)	24.4 (13.6)
2 Index*	4.8 (2.2)	6.1 (2.2)	21.7 (11.7)	5.6 (2.9)	5.3 (2.1)	25.1 (13.7)
3 Middle	4.8 (2.5)	5.4 (2.4)	24.0 (12.6)	4.2 (1.9)	6.5 (2.2)	21.3 (10.9)
4 Ring	4.3 (2.4)	5.2 (2.0)	19.1 (10.4)	3.7 (1.7)	5.2 (1.9)	19.5 (10.9)
5 Little	4.8 (1.9)	4.1 (1.6)	15.1 (8.0)	3.5 (1.6)	3.5 (2.2)	15.5 (8.5)

*Nonpreferred hand.

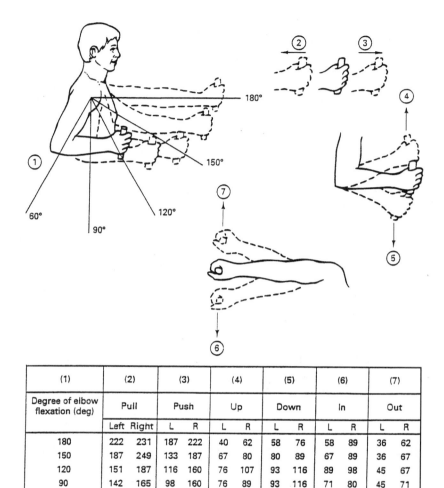

(1)	(2)		(3)		(4)		(5)		(6)		(7)	
Degree of elbow flexation (deg)	Pull		Push		Up		Down		In		Out	
	Left	Right	L	R	L	R	L	R	L	R	L	R
180	222	231	187	222	40	62	58	76	58	89	36	62
150	187	249	133	187	67	80	80	89	67	89	36	67
120	151	187	116	160	76	107	93	116	89	98	45	67
90	142	165	98	160	76	89	93	116	71	80	45	71
60	116	107	96	151	67	89	80	89	76	89	53	71

Figure 3-14. Fifth-percentile arm forces, in newtons, exerted by sitting men (adapted from U.S. Army Missile Command, MIL HDBK 759, 1981a.)

the pedals forward makes body weight less effective for generation of reaction force to the pedal effort; hence, a suitable backrest should be provided, against which the buttocks and low back press while the feet push forward on the pedal.

Small forces, such as for the operation of switches, can be generated in nearly all directions with the feet, but the downward or down-and-fore directions are preferred. The largest forces can be generated with extended or nearly extended legs; limited by body inertia in the downward direction, in the more forward direction both by inertia and the provision of buttock and back support surfaces. These prin-

Table 3-6. Mean poke forces (and standard deviations), in newtons, exerted by 30 subjects in the direction of the straight digits. (Reprinted with permission of the publisher from Kroemer, K. H. E., Kroemer, H. B., and Kroemer-Elbert, K. E. (1994). *Ergonomics: How to Design for Ease and Efficiency*. Englewood Cliffs, NJ: Prentice–Hall. All rights reserved.)

Digit	10 male mechanics	10 male students	10 female students
1. Thumb	83.8 (25.19) A	46.7 (29.19) C	32.4 (15.36) D
2. Index finger	60.4 (25.81) B	45.0 (29.99) C	25.4 (9.55) DE
3. Middle finger	55.9 (31.85) B	41.3 (21.55) C	21.5 (6.46) E

Note: Entries with different letters are significantly different from each other ($p \leq 0.05$).

ciples are typically applied in automobiles. For example, operation of a clutch or brake pedal can normally be performed easily with an almost right angle at the knee. But, if the power-assist system fails, very large forces must be exerted with the feet: in this case, thrusting one's back against a strong backrest and extending the legs are necessary to generate the needed pedal force.

Figures 3-15 through 3-19 provide information about the forces that can be applied to a pedal with legs and feet. Of course, the forces depend on body support and hip and knee angles. The largest forward thrust force can be exerted with the nearly extended legs, which leaves very little room to move the foot control further away from the hip.

The strength that can be exerted with the foot to an object such as a pedal depends on the joint strengths that can be transmitted along the chain ankle-knee-hip to the seat that provides the reaction support needed according to Newton's third law. This is charted in Figure 3-20. The foot exertion loads all proximal segments according to the prevailing angles and lever arms. Following the diagram outward, one sees that bad seat design; frail hip, knee, or ankle; or low friction at the coupling of the shoe with the object may all make for a weak kick.

Information on body strengths has been compiled, for example, by NASA (1989); the U.S. Army Missile Command (1981); Eastman-Kodak (1986); Kroemer, Kroemer and Kroemer-Elbert (1994); Salvendy (1987); Weimer (1993); and Woodson, Tillman, and Tillman (1991). However, caution is necessary when applying these data because they were measured on different populations under varying conditions.

SUMMARY

Muscle contraction is brought about by active shortening of muscle substructures. Elongation of the muscle is due to external forces.

Muscle contraction is controlled by excitation signals from the central nervous system. Each specific signal affects those fibers that are combined to a motor unit, of which there are many in a muscle.

Table 3-7. Mean forces (and standard deviations), in N, exerted by 21 male students and by 12 male machinists.

Couplings (See Figure 3-12)	Digit 1 (thumb)	Digit 2 (index)	Digit 3 (middle)	Digit 4 (ring)	Digit 5 (little)	
Push with digit tip in direction of the extended digit (Poke)	91 (39)* 138 (41)	52 (16)* 84 (35)	51 (20)* 86 (28)	35 (12)* 66 (22)	30 (10)* 52 (14)	See also Table 3-6
Digit Touch (Coupling #1) perpendicular to extended digit	84 (33)* 131 (42)	43 (14)* 70 (17)	36 (13)* 76 (20)	30 (13)* 57 (17)	25 (10)* 55 (16)	—
Same, but all fingers press on one bar	—	digits 2, 3, 4, 5 combined: 162 (33)				
Tip force (like in typing; angle between distal and proximal phalanges about 135 deg)	—	30 (12)* 65 (12)	29 (11)* 69 (22)	23 (9)* 50 (11)	19 (7)* 46 (14)	—
Palm Touch (Coupling #2) perpendicular to palm (arm, hand, digits extended and horizontal)	—	—	—	—	—	233 (65)
Hook Force exerted with digit tip pad (Coupling #3, Scratch)	61 (21) 118 (24)	49 (17) 89 (29)	48 (19) 104 (26)	38 (13) 77 (21)	34 (10) 66 (17)	all digits combined: 108 (39)* 252 (63)
Thumb-Fingertip Grip (Coupling #4, Tip Pinch)	—	1 on 2 50 (14)* 59 (15)	1 on 3 53 (14)* 63 (16)	1 on 4 38 (7)* 44 (12)	1 on 5 28 (7)* 30 (6)	—
Thumb-Finger Palmar Grip (Coupling #5, Pad Pinch)	1 on 2 & 3 85 (16)* 95 (19)	1 on 2 63 (12)* 34 (7)	1 on 3 61 (16)* 70 (15)	1 on 4 41 (12)* 54 (15)	1 on 5 31 (9)* 34 (7)	—
Thumb-Forefinger Side Grip (Coupling #6, Side Pinch)	—	1 on 2 98 (13)* 112 (16)	—	—	—	—
Power Grasp (Coupling #10, Grip Strength)	—	—	—	—	—	318 (61)* 366 (53)

Source: Reprinted with permission of the publisher from Kroemer, K.H.E., Kroemer, H.B., and Kroemer-Elbert, K.E. (1994). *Ergonomics: How to Design for Ease and Efficiency.* Englewood Cliffs, NJ: Prentice-Hall. All rights reserved.

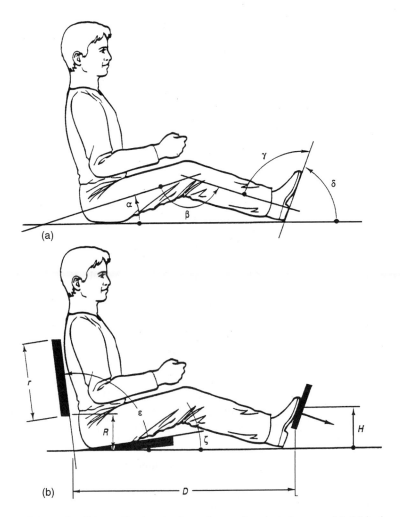

Figure 3-15. Conditions affecting the force that can be exerted on a pedal: (a) body angles and (b) work space dimensions. (reprinted with permission of the publisher from Kroemer, K.H.E., Kroemer, H.B., and Kroemer-Elbert, K.E. (1994). *Ergonomics: How to Design for Ease and Efficiency.* Englewood Cliffs, NJ: Prentice-Hall. All rights reserved.)

An efferent stimulus arriving from the CNS brings about a single-twitch contraction of the motor unit. A rapid sequence of stimuli can lead to a superposition of muscle twitches, which may fuse together into a sustained contraction called tetanus.

Prolonged strong contraction leads to muscular fatigue, which hinders the continuation of metabolic processes by reducing oxygen supply to, and metabolite

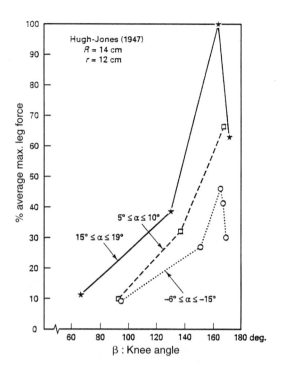

Figure 3-16. Effects of thigh angle α and knee angle β on pedal push force.
(Reprinted with permission of the publisher from Kroemer, K.H.E., Kroemer, H.B., and
Kroemer-Elbert, K.E. (1994). *Ergonomics: How to Design for Ease and Efficiency.*
Englewood Cliffs, NJ: Prentice-Hall. All rights reserved.)

removal from, the muscle. Hence, maximal voluntary contraction can be maintained for only a few seconds.

In isometric contraction, muscle length remains constant, which establishes a
static condition for the body segments affected by the muscle. In an isotonic effort, the muscle tension remains constant, which usually coincides with a static
(isometric) effort.

Dynamic activities result from changes in muscle length, which bring about
motion of body segments. In an isokinematic effort, speed remains unchanged. In
an isoinertial test, the mass properties remain constant.

Maximal muscle tension depends on the motivation of the person exerting
the effort, as well as on the individual's muscle size and exertion skill.

Human body (segment) strength is measured routinely as the force (or
torque) exerted to an instrument external to the body. Measurement of strength requires carefully controlled experimental conditions.

Design of equipment and work tasks for human body segment strength capabilities is done systematically by determining whether the exertion is static or dynamic; establishing with what body part the force or torque is exerted; following

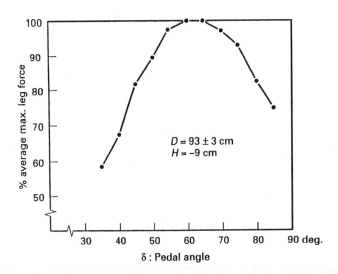

Figure 3-17. Effects of ankle (pedal) angle δ on foot force generated by ankle rotation. (Reprinted with permission of the publisher from Kroemer, K.H.E., Kroemer, H.B., and Kroemer-Elbert, K.E. (1994). *Ergonomics: How to Design for Ease and Efficiency.* Englewood Cliffs, NJ: Prentice-Hall. All rights reserved.)

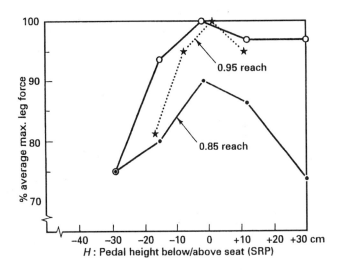

Figure 3-18. Effects of pedal height *H* and leg extension on pedal push force. (Reprinted with permission of the publisher from Kroemer, K.H.E., Kroemer, H.B., and Kroemer-Elbert, K.E. (1994). *Ergonomics: How to Design for Ease and Efficiency.* Englewood Cliffs, NJ: Prentice-Hall. All rights reserved.)

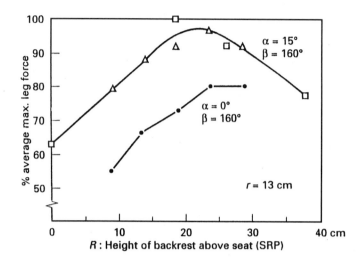

Figure 3-19. Effects of backrest height R on pedal push force. (Reprinted with permission of the publisher from Kroemer, K.H.E., Kroemer, H.B., and Kroemer-Elbert, K.E. (1994). *Ergonomics: How to Design for Ease and Efficiency.* Englewood Cliffs, NJ: Prentice-Hall. All rights reserved.)

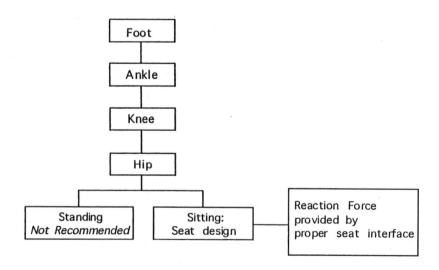

Figure 3-20. Determining the critical body segment strength for foot operation.

the chain of strength vectors through the involved body segments to find the weak link and to improve and rearrange, if possible; and selecting the body strength percentile (minimum and/or maximum) that is critical for the operation.

REFERENCES

Astrand, P.O., and Rodahl, K. (1977). *Textbook of Work Physiology,* 2nd ed. New York, NY: McGraw-Hill.

Astrand, P.O., and Rodahl, K. (1986). *Textbook of Work Physiology,* 3rd ed. New York, NY: McGraw-Hill.

Basmajian, J.V., and DeLuca, C.J. (1985). *Muscles Alive,* 5th ed. Baltimore, MD: Williams and Wilkins.

Best, C.H., and Taylor, N.B. (1973) *Physiological Basis of Medical Practice.* Baltimore, MD: Williams and Wilkins.

Buchthal, F., Guld, C., and Rosenfalck, P. (1957). Multielectrode Study of the Territory of a Motor Unit. *Acta Physiologica Scandinavica* 39:83–103.

Caldwell, L.S., Chaffin, D.B., Dukes-Dobos, F.N., Kroemer, K.H.E., Laubach, L.L., Snook, S.H., and Wasserman, D.E. (1974). A Proposed Standard Procedure for Static Muscle Strength Testing. *American Industrial Hygiene Association Journal* 35(4):201–206.

Chaffin, D.B., and Andersson, G.B.J. (1991). *Occupational Biomechanics,* 2nd ed. New York, NY: Wiley.

Eastman-Kodak Company (ed.) (1986). *Ergonomic Design for People at Work.* New York, NY: Van Nostrand Reinhold.

Enoka, R.M. (1988). *Neuromechanical Basis of Kinesiology.* Champaign IL: Human Kinetics Books.

Fukashiro, S., Komi, P.V., Jaervinen, M., and Miyashita, M. (1995). In Vivo Achilles Tendon Loading During Jumping. *European Journal of Applied Physiology* 71:453–458.

Kahn, J.F., and Monod, H. (1989). Fatigue Induced by Static Work. *Ergonomics* 32(7):839–846.

Komi, P.V., Belli, A., Huttunen, V., Bonnefoy, R., Geyssant, A., and Lacour, J.R. (1996). Optic Fibre as a Transducer of Tendomuscular Forces. *European Journal of Applied Physiology* 72:278–280.

Kroemer, K.H.E. (1986). Coupling the Hand with the Handle: An Improved Notation of Touch, Grip, and Grasp. *Human Factors* 28:337–339.

Kroemer, K.H.E. (1979). A New Model of Muscle Strength Regulation. In *Proceedings, Annual Conference of the Human Factors Society.* Santa Monica, CA: Human Factors Society, pp. 19–20.

Kroemer, K.H.E., Kroemer, H.B., and Kroemer-Elbert, K.E. (1994). *Ergonomics: How to Design for Ease and Efficiency.* Englewood Cliffs, NJ: Prentice-Hall.

Kroemer, K.H.E., and Marras, W.S. (1980). Toward an Objective Assessment of the Maximal Voluntary Contraction Component in Routine Muscle Strength Measurements. *European Journal of Applied Physiology* 45:1–9.

Kroemer, K.H.E., Marras, W.S., McGlothlin, J.D., McIntyre, D.R., and Nordin, M. (1989). *Assessing Human Dynamic Muscle Strength.* Technical Report, 8-30-89. Blacksburg, VA: Virginia Tech, Industrial Ergonomics Laboratory.

Also published (1990) in *International Journal of Industrial Ergonomics* 6:199–210.

NASA (1989). *Man-System Integration Standards* (Revision A). NASA-STD 3000A. Houston, TX: L.B.J. Space Center, SP 34-89-230.

Salvendy, G. (ed.) (1987). *Handbook of Human Factors and Ergonomics.* New York, NY: Wiley

Schneck, D.J. (1985). *Biomechanics of Striated Skeletal Muscle.* Santa Barbara, CA: Kinko.

Schneck, D.J. (1990). *Engineering Principles of Physiologic Function.* New York, NY: New York University Press.

Schneck, D.J. (1992). *Mechanics of Muscle,* 2nd ed. New York, NY: New York University Press.

Schuind, F., Garcia-Elias, M., Cooney, W.P., and An, K.N. (1992). Flexor Tendon Forces: In Vivo Measurements. *Journal of Hand Surgery* 17:291–298.

Soderberg, G.L. (ed.) (1992). *Selected Topics in Surface Electomyography for Use in the Occupational Setting: Expert Perspectives.* DDHS-NIOSH Publication 91-100. Washington, DC: Department of Health and Human Services.

U.S. Army Missile Command (1981a). MIL HDBK 759. *Human Factors Engineering Design for Army Material* (Metric). Philadelphia, PA: Naval Publications and Forms Center.

U.S. Army Missile Command (1981b). MIL STD 1472. *Human Engineering Design Criteria for Military Systems, Equipment, and Facilities.* Redstone Arsenal, AL: U.S. Army Missile Command.

Van Harreveld, A. (1947). On the Force and Size of Motor Units in the Rabbit's Sartorius Muscle. *American Journal of Physiology* 151:96–106.

Van Harreveld, A. (1948). The Structure of the Motor Units in Rabbit's *m. sartorius. Archiv Nierl. Physiol.* 28:408–412.

Weimer, J. (1993). *Handbook of Ergonomic and Human Factors Tables.* Englewood Cliffs, NJ: Prentice-Hall.

Winter, D.A. (1979). *Biomechanics and Motor Control of Human Behavior.* New York, NY: Wiley.

Woodson, W.E., Tillman, B., and Tillman, P. (1991). *Human Factors Design Handbook,* 2nd ed. New York, NY: McGraw-Hill.

FURTHER READING

Asimov, I. (1963). *The Human Body: Its Structure and Operation.* New York, NY: New American Library/Signet.

Astrand, P.O., and Rodahl, K. (1986). *Textbook of Work Physiology,* 3rd ed. New York, NY: McGraw-Hill.

Basmajian, J.V., and DeLuca, C.J. (1985). *Muscles Alive,* 5th ed. Baltimore, MD: Williams and Wilkins.

Chaffin, D.B., and Andersson, G.B.J. (1991). *Occupational Biomechanics,* 2nd ed. New York, NY: Wiley.

Daniels, L., and Worthingham, C. 1980. *Muscle Testing,* 4th ed. Philadelphia, PA: Saunders.

Enoka, R.M. (1988). *Neuromechanical Basis of Kinesiology.* Champaign IL: Human Kinetics Books.

Kroemer, K.H.E. (1970). Human Strength: Terminology, Measurement and Interpretation of Data. *Human Factors* 12:279–313.

Kroemer, K.H.E., Kroemer, H.B., and Kroemer-Elbert, K.E. (1994). *Ergonomics: How to Design for Ease and Efficiency.* Englewood Cliffs, NJ: Prentice-Hall

Kroemer, K.H.E., and Marras, W.S. (1980). Toward an Objective Assessment of the Maximal Voluntary Contraction Component in Routine Muscle Strength Measurements. *European Journal of Applied Physiology* 45:1–9.

Radwin, R.C., Beebe, D.J., Webster, J.G., and Yen, T.Y. (1996). Instrumentation for Occupational Ergonomics. Chapter 7 in A. Bhattacharia and J.D. McGlothlin (eds.). *Occupational Ergonomics. Theory and Applications.* New York, NY: Marcel Dekker, pp. 165–193.

Schmidt, R.F. and Thews, G. (eds.) (1989). *Human Physiology* (2nd ed.). Berlin, Germany; New York, NY: Springer.

Winter, D.A. (1990) *Biomechanics and Motor Control of Human Movement,* 2nd ed. New York, NY: Wiley.

Glossary of Muscle Terms

acceleration   Second time derivative of displacement.

actin   muscle filament (see there) capable of sliding along myosin (see there).

action   Activation of muscle; see *contraction.*

body (segment) strength   Ability to exert force or torque to an object touching the body.

cocontraction   Simultaneous contraction of two or more muscles.

concentric (muscle effort)   Shortening of a muscle against a resistance.

contraction   Literally, "pulling together" the z-lines delineating the length of a sarcomere, caused by the sliding action of actin and myosin filaments. Contraction develops muscle tension only if the shortening is resisted. (Note that, during an isometric "contraction," no change in sarcomere length occurs and, in an eccentric "contraction," the sarcomere is actually lengthened. To avoid such contradiction in terms, it is often better to use the terms *activation, effort,* or *exertion.*)

displacement   Distance moved (in a given time).

distal   Away from the center of the body.

dynamics   A subdivision of mechanics that deals with forces and bodies in motion.

eccentric (muscle effort)   Lengthening of a resisting muscle by external force.

effort (of muscle)   See *contraction.*

exertion (of muscle)   See *contraction.*

fiber See *muscle.*

fibril See *muscle fibers.*

filament See *muscle fibers.*

force According to Newton's third law, the product of mass and acceleration. The proper unit is the newton, with $1 \text{ N} = 1 \text{ kg m/s}^2$. On earth, 1 kg applies a (weight) force of 9.81 N (1 lb exerts 4.448 N) to its support. Muscular force is defined as tension multiplied by the transmitting cross-sectional area.

free dynamic In this context, an experimental condition in which neither force nor displacement and its time derivatives are manipulated as independent variables.

isoacceleration A condition in which the acceleration is kept constant.

isoforce A condition in which the muscular force (tension) is constant, that is, isokinetic. This term is equivalent to *isotonic.*

isoinertial A condition in which muscle moves a constant mass.

isojerk A condition in which the time derivative of acceleration, jerk, is kept constant.

isokinematic A condition in which the velocity of muscle shortening (or lengthening) is constant. (Depending on the given biomechanical conditions, this may or may not coincide with a constant angular speed of a body segment about its articulation.) Compare with *isokinetic.*

isokinetic A condition in which muscle tension (force) is kept constant. See *isoforce* and *isotonic;* compare with *isokinematic.*

isometric A condition in which the length of the muscle remains constant.

isotonic A condition in which muscle tension (force) is kept constant; see *isoforce.* (Occasionally in the past, this term was falsely applied to any condition other than isometric.)

jerk Third time derivative of displacement.

kinematics A subdivision of dynamics that deals with the motions of bodies but not the causing forces.

kinetics A subdivision of dynamics that deals with forces applied to masses.

mechanical advantage In this context, the lever arm (moment arm, leverage) at which a muscle works around a bony articulation.

mechanics The branch of physics that deals with forces applied to bodies and their ensuing motions.

moment The product of force and the length of the (perpendicular) lever arm at which it acts; mechanically equivalent to torque.

motor unit   All muscle filaments under the control of one efferent nerve axon.

muscle   A bundle of fibers, able to contract or be lengthened; in this context, striated (skeletal) muscle that moves body segments about each other under voluntary control.

muscle contraction   The result of contractions of motor units distributed through a muscle so that the muscle length is shortened.

muscle fibers   Elements of muscle, containing fibrils that consist of filaments.

muscle fibrils   Elements of muscle fibers, containing filaments.

muscle filaments   Muscle fibril elements (actin and myosin, polymerized protein molecules) capable of sliding along each other, thus shortening the muscle and, if doing so against resistance, generating tension.

muscle force   The product of tension within a muscle multiplied with the transmitting muscle cross section.

muscle strength   The ability of a muscle to generate and transmit tension in the direction of its fibers. See also *body strength*.

muscle tension   The pull within a muscle expressed as force divided by transmitting cross section.

myo-   Prefix referring to muscle (Greek *mys*, "muscle").

myosin   muscle filament (see there) along which actin (see there) can slide.

mys-   Prefix referring to muscle (Greek *mys*, "muscle").

proximal   Toward the center of the body.

rate coding   The time sequence in which efferent signals arrive at a specific motor unit and cause it to contract.

recruitment coding   The time sequence in which efferent signals arrive at different motor units and cause them to contract.

repetition   Performing the same activity more than once. (One repetition indicates two exertions.)

rhythmic   The same action repeated in regular intervals.

statics   A subdivision of mechanics that deals with bodies at rest.

strength   See *body strength* and *muscle strength*.

tension   Force divided by the cross-sectional area through which it is transmitted.

torque   The product of force and the length of the (perpendicular) lever arm at which it acts; mechanically equivalent to moment.

velocity   First time derivative of displacement.

Chapter 4

The Neuromuscular Control System

OVERVIEW

The central nervous system is one of several control and regulation systems of the body. It collects inputs from various sensors that respond to internal and external stimuli. Its integration and regulation functions concerning motor activities are mainly in the cerebrum, the cerebellum, and the spinal cord. The pathways for incoming and outgoing signals are the neurons, which possess the ability to inhibit or facilitate the transmission of impulses.

THE MODEL

The nervous system transmits (feeds back) information about events inside and outside the body from various sensors along its afferent pathways to the brain. Here, decisions about appropriate actions and reactions are made, and (feedforward) signals are generated and sent along the efferent pathways to the muscles.

INTRODUCTION

The purpose of the human regulatory and control systems is to maintain equilibrium (homeostasis) on the cellular level and throughout the body despite changes in the strain on the body generated by varying external environments and work requirements. A detailed consideration of the several parts and various mechanisms of this control system is challenging and stimulating but exceeds the scope of this book. This text considers primarily the neuromuscular system, which is rather well researched.

The human body is under the control of a dual system: both the hormonal (endocrine) system and the nervous system have similar functions. Within the hormonal system, one set of hormones (norepinephrine) stimulates smooth muscle in some organs but, in others, inhibits the muscular contractions; another hormone (acetylcholine) has just the opposite effect on the same smooth muscles. Not much is known about the exact locations in the brain at which functions of the autonomic nervous system are regulated. This chapter concentrates on the nervous system, particularly as it affects functions of skeletal muscle.

ORGANIZATION OF THE NERVOUS SYSTEM

In terms of function, there are two subdivisions of the nervous system: the autonomic (visceral) system and the somatic system. The *autonomic* system generates "fright, flight, or fight" attitudes and regulates involuntary functions, such as cardiac and smooth muscle, blood flow, digestion, or glucose release in the liver. It is subdivided into the sympathetic and the parasympathetic subsystems, which control mental activities, conscious actions, and the skeletal muscles. The *somatic* (from the Greek *soma,* "body") nervous system regulates mental activities, skeletal muscle, and conscious actions.

Anatomically, the nervous system is divided into three major subdivisions: The *central* nervous system (CNS) includes brain and spinal cord; it has primarily control functions. The *peripheral* nervous system (PNS) includes the cranial and spinal nerves; it does not control functions but transmits signals to and from the brain. The third is the *autonomic* nervous system just described.

Central Nervous System

As Figure 4-1 shows, the brain is usually divided into several sections. Of particular interest for the neuromuscular control system is the forebrain's *cerebrum,*

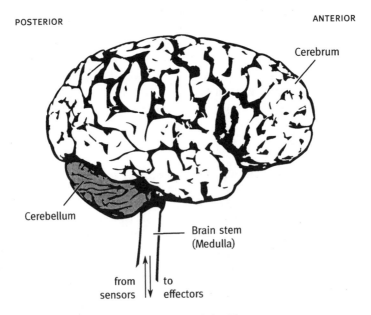

Figure 4-1. The human brain seen from the right side.

which consists of the two (left and right) cerebral hemispheres, each divided into four lobes. Voluntary movements are controlled, and sensory experience, abstract thought, memory, learning, and consciousness are located in the cerebrum. The *cortex,* which is the many-folded top layer of the cerebrum, contains the motor cortex, which controls voluntary movements of the skeletal muscles, and the sensory cortex, which interprets sensory inputs.

The basal ganglia of the midbrain are composed of large pools of neurons, which control semivoluntary complex activities such as walking. Part of the hindbrain is the *cerebellum,* which distributes and integrates impulses from the cerebral association centers to the motor neurons in the spinal cord. The *brain stem* controls the functions of such vital organs as the heart and lungs.

The *spinal cord* is an extension of the brain. From the uppermost section of the spinal cord, 12 pairs of cranial nerves emanate; below, 31 pairs of spinal nerves have their roots between the vertebrae. All serve defined sectors of the body. Nerves are both sensory and motor pathways, carrying signals from sensors to the spinal cord and also from the cord to the organs, including the muscles.

Sensors of the Peripheral Nervous System

The central nervous system, CNS, receives information arriving from the sensory part of the peripheral nervous system, PNS. (This sensory part is also called feedback or afferent section of the PNS.) It carries signals concerning the outside from *external* receptors (exteroceptors), which respond to light, sound, touch, temperature, and chemicals, and from *internal* receptors (interoceptors), which report changes within the body. Since all these sensations come from various parts of the body, external and internal receptors together are also called somaesthetic sensors.

Internal receptors include the *proprioceptors.* Among these are the muscle spindles, which are nerve filaments wrapped around small muscle fibers; they detect the amount of stretch of the muscle. Golgi organs are associated with muscle tendons and detect their tension; hence, they report to the central nervous system information about the strength of contraction of the muscle. Finally, Ruffini organs are kinesthetic receptors located in the capsules of articulations. They respond to the degree of angulation of the joints (joint position) and to change in general, as well as to the rate of change.

The sensors in the *vestibulum* are also proprioceptors that detect and report the position of the head in space and respond to sudden changes in its attitude. This is done by sensors in the semicircular canals, of which there are three, each located in another orthogonal plane. To relate the position of the body to that of the head, proprioceptors in the neck are triggered by displacements between trunk and head.

Another set of interoceptors, called visceroceptors, report on the events within the visceral (internal) structures of the body, such as organs of the abdomen and chest, as well as on events in the head and other deep structures. The usual modalities of visceral sensations are pain, burning sensations, and pressure. Since the same sensations are also provided from external receptors and since the pathways of visceral and external receptors are closely related, information about the body is often integrated with information about the outside.

External receptors provide information about the interaction between the body and the outside: sight (vision), sound (audition), taste (gustation), smell (olfaction), temperature, chemical agents, and touch (taction). Several of these are of particular importance for the control of muscular activities: the sensations of touch, pressure, and pain can be used as feedback to the body regarding the direction and intensity of muscular activities transmitted to an outside object (see Chap. 3). Free nerve endings, Meissner's and Pacinian corpuscles, and other receptors are located throughout the skin of the body, however in different densities. They transmit the sensations of touch, pressure, and pain. Since the nerve pathways from the free endings interconnect extensively, the sensations reported are not always specific for a modality; for example, very hot or cold sensations can be associated with pain, which may also be caused by hard pressure on the skin.

Almost all sensors respond vigorously to a change in the stimulus but will report less and less in the next seconds or minutes if the load stays constant. This adaptation makes it possible to live with, for example, the continued pressure of clothing. The speed of adaptation varies with the sensors. Furthermore, the speeds with which the sensations are transmitted to the central nervous system are quite different for different sensors: light and sound, for example, cause the fastest reactions, and pain often the slowest.

Effectors of the Peripheral Nervous System

Decisions made in the CNS are carried as action signals along the motor part of the PNS. (It is also called the feedforward or efferent section of the PNS.) Their main effects (as pertain to this book) are to make muscle motor units contract (see Chap. 3).

THE NERVOUS PATHWAYS

Information from a sensor is passed along an afferent path to the decision maker in the spinal column or brain. In many cases, this results in a signal sent along the efferent pathway to the effecting organ, often a muscle.

The spinal cord provides the peripheral nerve network to and from the brain. It is contained within the protective bone structures of the spinal column, passing through the openings of the vertebral foramen; see Chapter 2. At the intersections of the vertebrae, nerves pass horizontally between the bones and extend into body tissues, where they deliver and pick up signals. This network of motor nerves resembles the lower part of a tree; with respect to its sensory function, it is often explained as (nerve) "roots" that come from the tissues and feed at the vertebrae into the main "trunk" of the spinal cord.

The uppermost section of the spinal cord contains the 12 pairs of *cranial* nerves, which serve structures in the head and neck as well as the lungs, heart, pharynx and larynx, and many abdominal organs. These cranial nerves control eye, tongue, and facial movements and the secretion of tears and saliva. The main inputs are from the eyes, the taste buds in the mouth, the nasal olfactory receptors, and touch, pain, heat, and cold receptors of the head.

Below the neck, 31 pairs of *spinal* nerves emanate between the thoracic and lumbar vertebrae and serve defined sectors of the rest of the body. Figure 4-2 shows the origins of the spinal nerves in the sections of the spinal column and the defined areas of the skin (dermatomes) that they innervate.

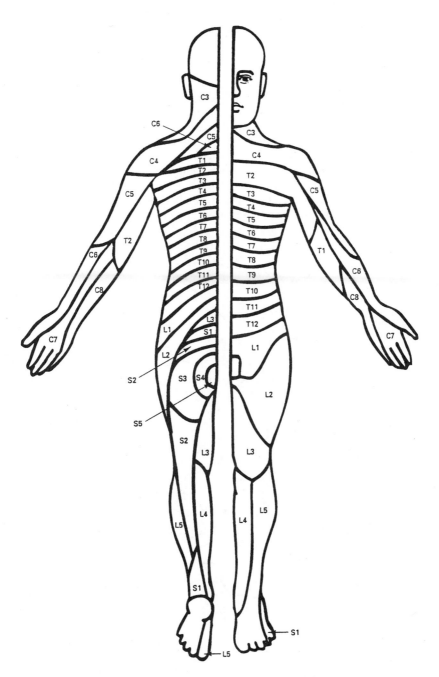

Figure 4-2. Sensory dermatomes with their spinal nerve roots: *C* stands for cervical, *T* for thoracic, *L* for lumbar, and *S* for sacral.

The nerves are mixed sensory and motor pathways, carrying both somatic and autonomic signals between the brain, spinal cord, and organs throughout the body.

The Neuron

The basic functional unit of the nervous transmission system is the nerve cell, called the *neuron*. Neurons transmit signals from one to another through their filamentous nerve fibers. In the brain, neurons are responsible for the storage of memories, for patterns of thinking, for initiating motor responses, and for many other capabilities. There are about 12 billion neurons in the brain and spinal cord. A *synapse* is the junction between two neurons through which signals are transmitted. The synapse has a switching ability; that is, it may or may not transmit signals.

Figure 4-3 sketches a typical neuron, combining the main features of motor neurons, sensory neurons, and interneurons. It consists of three major parts: the

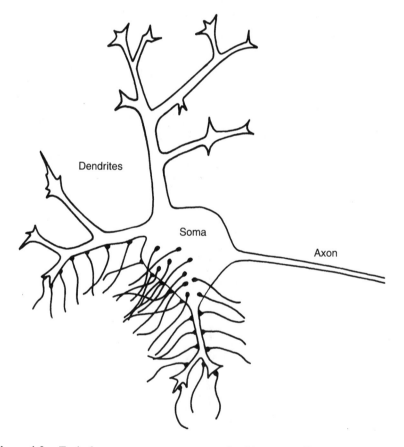

Figure 4-3. Typical neuron components: soma, dendrites, axon. Snyapses are sketched only in the lower left part of the figure.

main body (soma), and the processes, either a long extension, the axon, or the branching dendrite.

Each motor neuron has only one *axon,* which serves to transmit signals from the cell body. At a distance from the soma, the axon branches out into terminal fibrils that connect with other neurons. The length of the axon may be only a few millimeters or a meter or longer. The neuron has up to several hundred *dendrites,* projections of usually only a few millimeters, which receive signals from the axons of other neurons and transmit these into their own neuron cell body. Shown (only in the lower left corner of Fig. 4-2) are synapses, the end points of (hundreds of) fibrils coming from axons of other neurons. Synapses are formed like knobs, bulbs, clubs, or feet. Each synaptic body has numerous vesicles, which contain transmitter substance. The synaptic membrane is separated from the opposing subsynaptic membrane of the neuron by a synaptic cleft, a space of about 2 Å.

Transmission of Nerve Signals

Some axon terminals secrete an excitatory transmitter substance, while others carry an inhibitory neurohumor; this means that some terminals excite the neuron and others inhibit it. Among the excitatory neurohumors is norepinephrine, while one of the inhibitory secretions is dopamine. Certain chemical agents (such as some anesthetics and curare) can prevent the secretion of excitatory neurohumor, thus inhibiting the transmission of signals and possibly causing paralysis of the musculature.

Excretion of an excitatory transmitter increases the permeability of the subsynaptic membrane beneath the synaptic knob, which allows sodium ions to flow rapidly to the inside of the cell. Since sodium ions carry positive charges, the result is an increase in positive charge inside the cell, bringing about an excitatory postsynaptic potential (see the following section, "The Sodium Pump"). This sets up an electrical current throughout the cell body and its membrane surfaces, including the base of the axon. If this potential becomes high enough, it initiates an *action potential* in the axon. If the potential does not exceed a threshold value, no action impulse will be transmitted along the axon. Usually, the threshold is exceeded by a summation of inputs. Such summation can result either from simultaneous firing of several synapses or from rapid repeated firing of one synapse. If the threshold is exceeded, the axon carries repeated action impulses as long as the potential remains above the threshold. The higher the postsynaptic potential rises, the more rapidly the axon fires. Motor signal intensities are typically within the range of ±70 mV along the axon.

Depending on their type and size, some neurons are able to transmit as many as a thousand impulses per second along their axons, while others may not be able to transmit more than 25 signals per second.

The inhibitory transmitters have an opposite effect on the synapse: they create a negative potential called the inhibitory postsynaptic potential. This reduces the actual synaptic potential, which may result in a value below the threshold of the neuron: different neurons have different thresholds. Thus, the inhibitory synaptic knobs may stop or prevent neuron discharge.

Most nervous impulses are not carried by a single neuron from receptor to destination but follow a chain of linked neurons. The transmission can be interrupted if it reaches a neuron with a particularly high threshold. Once developed, the action potential propagates itself point by point along a nerve fiber without needing further stimulation. At each successive point along the fiber, the action potential rises to a maximum and then rapidly declines. Because of this shape, the action potential is occasionally called spike potential. The height of its spike, its strength, does not vary with the strength of the stimulus, nor does it weaken as it progresses along the axon. Either the stimulus is strong enough to elicit a response or there is no potential; this is an all-or-none phenomenon similar to the one discussed in the contraction of skeletal muscle.

Although nerve fibers fatigue hardly at all, synapses do fatigue, some very rapidly, some rather slowly.

The velocity of the nerve impulse is a constant for each nerve fiber, ranging from 0.5 m/s to about 150 m/s. This speed does not change along a particular fiber and is correlated with the diameter of the fiber, being faster in a thick fiber than in a thin one. In the peripheral parts of the nervous systems, axons and dendrites are sheathed along their length. The envelope consists of myelin, a white material composed of protein and phospholipid. The presence of a myelin sheath allows a larger speed of conduction. Skeletal muscles are served by thick myelinated axons terminating at the motor end plates, while pain fibers are the thinnest and nonmyelinated.

Nerve fibers are extensions from a single neuron, hence, parts of a single cell. Nerves, in contrast, are multicellular structures, bundles of nerve fibers gathered from many neurons and arranged somewhat like the wires within a cable.

The Sodium Pump

In an excited neuron, the potential change is brought about by a redistribution of the positive sodium (Na) and potassium (K) ions on either side of the cell membrane. With a change of permeability of the membrane, sodium ions rush inward and potassium ions move out of the fiber. However, the positive Na ions overshoot an equilibrium point and become more concentrated inside the fiber than in the interstitial fluid. This causes the interior of the fiber to become positively charged and the external surface to become negative, that is, less positive. At this moment, the polarity of the membrane has been established as a change from its state of an undisturbed cell. When the positive K ions move out, the membrane becomes repolarized, reestablishing a positive charge on its external surface and leaving the cell interior less positive, that is, more negative. When the action potential moves on to the neighboring region on the fiber, this "sodium (and potassium) pump" restores the original ionic balance, with the positive Na ions moved out and the positive K ions recaptured.

Reflexes

The spinal cord is a center of coordination for certain actions, particularly limb movements, that do not need to involve the brain proper. Such a reflex usually begins with a stimulation of a peripheral sensory receptor. Its signal is sent as afferent impulse to the spinal cord, where it evokes a quick response, which is sent as efferent signal to the appropriate muscles. In this way, a reactive action of an effector muscle can be executed a few milliseconds after the stimulus was received since no time-consuming higher brain functions are involved. All effectors are either muscle fibers or gland cells; hence, the result of a reflex is either a muscular contraction or a gland secretion.

Control of Muscle Movements

While some motor functions of the body can be performed without involvement of higher brain centers, many complex voluntary muscular activities need fine regulation. These require various degrees of involvement of the higher brain centers, such as the cerebral cortex, the basal ganglia, and the cerebellum. It is believed that the motor cortex controls mostly very fine discrete muscle movements and that the basal ganglia have large pools of neurons organized for the control of complex movements, such as walking, running, and posture control. (This may be the locus for the "executive programs" mentioned in Chap. 3.) The motor pathway is the location at which efferent signals can be picked up by wire, needle, or surface electrodes and recorded by their electrical activities in an *electromyogram* (EMG). A suitable location for electrode placement is near the motor end plates of the innervated motor unit of the muscle of interest (see Chap. 3).

Avoiding Control Impairment
Through Ergonomic Engineering

Impairment of a motor nerve reduces the ability to transmit signals to innervated muscle motor units. Thus, motor nerve impairment impedes the controlled activity of muscles and, hence, reduces the ability to generate force or torque for application to tools, equipment, and work objects. Sensory nerve impairment reduces the information that can be brought back from sensors to the central nervous system. Sensory feedback is very important for many activities because it contains information about force and pressure applied, position assumed, and motion experienced. Sensory nerve impairment usually brings about sensations of numbness, tingling, or even pain. The ability to distinguish hot from cold may be reduced. Impairment of an autonomic nerve reduces the ability to control such functions as sweat production in the skin. A common sign of autonomic nerve impairment is dryness and shininess of skin areas controlled by that nerve.

Probably the most severe impairment of nervous control is caused by spinal cord injuries, frequent with horseback riding and automobile accidents in which the head is suddenly displaced with respect to the trunk. Too often, this results in a dislocation or breakage of vertebrae which, in turn, damages the spinal cord passing through the vertebral foramen (see Chap. 2). Damage to the cord may render it incapable of transmission of feedforward signals to body areas innervated by nerves below the injury point and also makes it unable to conduct feedback signals from these body parts to the brain. The victim has no control over, or sensation in, these body areas.

Nerve functioning can be affected also by repeated or sustained pressure. Such pressure may stem from bones, ligaments, tendons, sheaths, and muscles within the body or from hard surfaces and sharp edges of workplaces, tools, and equipment. Pressure within the body can occur if the position of a body segment reduces the passage opening through which a nerve runs. Another source of compression, or an added one, may be irritation and swelling of other structures within this opening, often of tendons and tendon sheaths. Carpal tunnel syndrome is a typical case of nerve compression (Pfeffer et al. 1988).

Carpal tunnel syndrome (CTS) is among the best-known cumulative trauma disorders. It was first mentioned in the literature in the mid-1900s. In 1959, Tanzer described several CTS cases: Two of his patients had recently started to milk cows on a dairy farm; three worked in a shop handling objects on a conveyor belt; two had been gardening, with considerable hand weeding; one had been using a spray gun with a finger trigger. Two patients had been working in large kitchens, where they stirred and ladled soup twice daily for about 600 students.

In 1966 and 1972, Phalen published his "classical" reviews: he described the typical gradual onset of numbness in the thumb and the first two and a half fingers of the hand (supplied by the median nerve), with the little finger and the ulnar side of the ring finger unaffected. In 1975, Birkbeck and Beer described the results of a survey they made of the work and hobby activities of 658 patients who suffered from CTS. Four out of five patients were employed in work requiring light, highly repetitive movements of the wrists and fingers. In 1976, Posch and Marcotte analyzed 1,201 cases of CTS. (Publications by Putz-Andersson, 1988; Kroemer, 1992; and Kuorinka and Forcier, 1995, provide more details on cumulative trauma disorders and on older studies.)

Yet, CTS is not only caused or aggravated by "heavy" activities at work and leisure activities (such as gardening and piano playing) but is also frequent with repetitive "light" work in the office or at home. In 1964, Kroemer described the occurrence of CTDs in typists and their possible relation to key force and displacement, key operation frequency, and posture of arms and hands. In 1980, Huenting, Grandjean, and Maeda frequently found impairments in the hands and arms of operators of accounting machines. In the 1980s and 1990s, an epidemic of CTS complaints by computer keyboard operators has been keeping health care providers and lawyers busy (Kroemer, 1996, 1997).

Anatomical and biomechanical knowledge helps us understand the problem. On the palmar side of the wrist, near the base of the thumb, the carpal bones form a concave floor. This "canal" is covered by ligaments, of which the transverse ligament is firmly fused to the carpal bones on the left and right. This structure is called the carpal tunnel. As Figure 4-4 shows, through its opening pass all the flexor tendons of the digits and the median nerve and blood vessels.

The median nerve innervates the thumb, much of the palm, and the index and middle fingers, as well as the radial side of the ring finger. Outside pressure on the tunnel ligament reduces the available passage space, as does a change from the straight wrist posture. Swelling of tissues inside the carpal tunnel, often of tendons and their sheaths (due to mechanical overexertion through often repeated digit motions, as in piano playing or other keyboarding), reduces the available passage space, and all these can generate pressure on the blood vessels and on the median nerve. This impairs blood flow and nerve functioning, which may at first be felt as occasional tingling or numbness; then pain can develop, together with

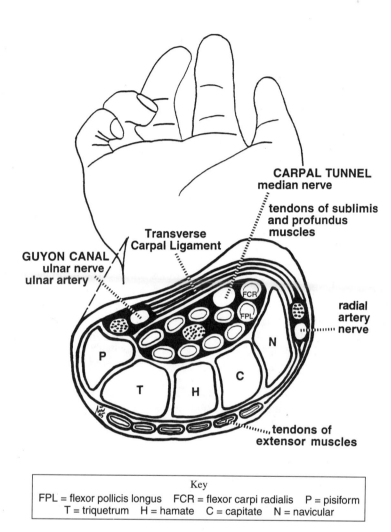

Figure 4-4. Cross section of the hand at the base of the thumb showing the "carpal tunnel" through which pass the flexor tendons of the thumb and fingers, the median nerve, and blood vessels.

increasing difficulty in performing hand activities. The medical diagnosis of CTS is often based on the measurement of reduced conduction velocity in the afflicted median nerve.

What can the ergonomist do to prevent the occurrence of such injuries? The violent trauma often associated with a car accident can be avoided by keeping the body parts properly aligned, such as by a large, sturdy backrest (with headrest) of the seat that fully supports the body in the case of a blow from behind, and by restraint systems with seat belts and air bags that prevent whiplash and impact on car structures.

In the case of repetitive activities where many minitraumas sum up to an injury, both the nature of the actions and their number of recurrences can be controlled. For example, inserting and hammering during assembly can be reduced in intensity and frequency by using other tools or work processes, such as gluing; or entering text can be done by voice recognition rather than by manual keying. It is most important to assure that the nature of the job "fits the human." A secondary improvement is to reduce the task duration and repetition rate by inserting other activities, or by employing relief workers or, sometimes, by simply taking a break. The ergonomics/human factors literature provides further guidance.

SUMMARY

The body must continuously control muscle functions according to information about conditions and events reported from various sensors. The information feedback to the central nervous system (where decisions are made and actions initiated) and the feedforward signals to the muscles flow along the peripheral nervous system.

Afferent (feedback) and efferent (feedforward) signals are transmitted along neurons, consisting of soma, dendrites, and axon. At the neuron, nerve fibrils (from other neurons) end in synapses that serve as selective switches. Depending on the strength of the incoming signal, it either is or is not transmitted across the synaptic membrane.

With many sensors reacting to the same stimulus and many afferent pathways transmitting the signals at different intensities and speeds, the peripheral nervous system serves as a filter or selector for the central nervous system.

REFERENCES AND FURTHER READING

Astrand, P.O., and Rodahl, K. (1986). *Textbook of Work Physiology,* 3rd ed. New York, NY: McGraw-Hill.

Birkbeck, M.Q., and Beer, T.C. (1975). Occupation in Relation to the Carpal Tunnel Syndrome. *Rheumatology and Rehabilitation* 14:218–221.

Guyton, A.C. (1979). *Physiology of the Human Body,* 5th ed. Philadelphia, PA: Saunders.

Huenting, H., Grandjean, E. and Maeda, K. (1980). Constrained Posture in Accounting Machine Operators. *Applied Ergonomics* 11(3): 145–149.

Kroemer, K.H.E. (1964). Über den Einfluss der räumlichen Lage von Tastenfeldern auf die Leistung an Schreibmaschinen. *Intern. Zeitschrift Angewandte Physiologie einschl. Arbeitsphys.* 20: 240–251.

Kroemer, K.H.E. (1992). Avoiding Cumulative Trauma Disorders in Shop and Office. *Journal of the American Industrial Hygiene Association* 53 (9):596–604.

Kroemer, K.H.E. (1996). The Role of the Ergonomics Expert in Personal Injury Cases. Chap. 6 in *1996 Wiley Expert Witness Update—New Developments in Personal Injury Litigation.* New York, NY: Wiley, pp. 167–184.

Kroemer, K.H.E. (1997). *Reviews of Publications Related to Keyboarding—in Chronological Order* (Report, 5 April 1997). Radford, VA: K.H.E. Kroemer Ergonomics Research Institute.

Kroemer, K.H.E., Kroemer, H.B., and Kroemer-Elbert, K.E. (1994). *Ergonomics: How to Design for Ease and Efficiency.* Englewood Cliffs, NJ: Prentice-Hall.

Kuorinka, I., and Forcier, L. (eds.) (1995). *Work Related Musculoskeletal Disorders: A Reference Book for Prevention.* London, UK: Taylor & Francis.

Pfeffer, G.B., Gelberman, R.H., Boyes, J.H., and Rydevic, B. (1988). The History of Carpal Tunnel Syndrome. *Journal of Hand Surgery* 13B(1): 28–34.

Phalen, G.S. (1966). The Carpal-Tunnel Syndrome—Seventeen Years' Experience in Diagnosis and Treatment of Six-Hundred Fifty-Four Hands. *Journal of Bone and Joint Surgery* 48A(2):211–228.

Phalen, G.S. (1972). The Carpal-Tunnel Syndrome. Clinical Evaluation of 598 Hands. *Clinical Orthopaedics and Related Research* 83 (March-April Issue): 29–40.

Plog, B.A. (ed.). (1996). *Fundamentals of Industrial Hygiene,* 4th ed. Itasca, IL: National Safety Council.

Posch, J.L., and Marcotte, D.R. (1976). Carpal Tunnel Syndrome—An Analysis of 1,201 Cases. *Orthopaedic Review* 5(5):24–35.

Putz-Anderson, V. (1988). *Cumulative Trauma Disorders: A Manual for Musculoskeletal Diseases of the Upper Limbs.* London, UK: Taylor & Francis.

Tanzer, R.C. (1959). The Carpal-Tunnel Syndrome—A Clinical and Anatomical Study. *American Journal of Industrial Medicine* 11:343–358.

Tanzer, R.C. (1959). The Carpal-Tunnel Syndrome—A Clinical and Anatomical Study. *Journal of Bone and Joint Surgery* 41A (4):626–634.

Weller, H., and Wiley, R.L. (1979). *Basic Human Physiology.* New York, NY: Van Nostrand Reinhold.

Chapter 5

Biomechanics

OVERVIEW

Biomechanics attempts to explain characteristics and responses of the human body as a biological system in mechanical terms. Biomechanical approaches have developed from research procedures and techniques stemming from physics (mechanics), mathematics (including computer sciences), anatomy, orthopedics, physiology and, particularly, anthropometry. In this chapter, biomechanical procedures are used to explore and explain the action of muscles bridging body joints, to assess mass properties of the human body and the transmission of forces and torques along a chain of body segments. Precise terminology to describe human body position and motion is discussed.

THE MODEL

The body is built on and around a skeleton of links (the long bones) that is articulated at the joints and powered by muscles. The body segments possess mass and other physical, material properties that are accounted for using mechanical principles.

INTRODUCTION

Biomechanics is not a new science. Leonardo da Vinci (1452–1519) and Giovanni Alfonso Borelli (1608–1679) combined physics with anatomy and physiology. In his book *De Motu Animalium (The Motion of Animals),* Borelli developed a model of the human skeleton consisting of a series of links (long bones) joined at their articulations and powered by muscles bridging the articulations. This "stick person" approach still underlies many current biomechanical models of the human body.

Development of the biomechanical sciences is closely linked to the physical laws developed by Newton (1642–1727) and achieved a first high point in the late 1800s when Harless determined the masses of body segments, Braune and Fischer investigated the interactions between mass distribution and external impulses applied to the human body, and von Meyer discussed statics and mechanics of the human body. (See Kroemer et al., 1988, for more details.) Since then, biomechanical research has addressed, among other topics, the response of the human body to vibrations and impacts, human strength and motion regarding the whole body or specific segments thereof, functions of the spinal column, hemodynamics and the cardiovascular system, and prosthetic devices (King and Chou, 1976; King, 1984).

Despite the large body of information available, biomechanics is still a developing scientific and engineering field with a wide variety of focus points, research methods, and measurement techniques for producing new theoretical and practical results. While using the substantial existent data and knowledge base, the practicing engineer must studiously follow the progress reported in the scientific and engineering literature in order to stay abreast of developments so that the most appropriate information can be applied to the design and management of human/machine systems. At the end of this chapter there is a partial list of journals in which much biomechanical information is published.

Treating the human body as a mechanical system—disregarding mental functions, for example—leads to many and gross simplifications. Furthermore, many components of the body are considered solely in mechanical terms, such as:

- Bones: lever arms, central axes, structural members
- Articulations: joints and bearing surfaces, often considered frictionless
- Articulation linings: lubricants, joint structures
- Tendons: cables transmitting muscle forces
- Tendon sheaths: pulleys and sliding surfaces, often considered frictionless
- Anthropometric data: dimensions of body segments, both in their surfaces and internally, obtained from statistical manipulation of data measured from groups of people
- Flesh: volumes, masses
- Contours: surfaces of geometric bodies
- Nerves: control and feedback circuits
- Muscles: motors, dampers, or locks
- Tissue: elastic load-bearing surfaces, springs, and contours
- Organs: generators or consumers of energy

This list briefly indicates some of the limitations imposed by biomechanical considerations and simplifications in modeling the human body.

Stress and Strain

In engineering terms, *strain* is the result or effect of *stress:* stress is the input, strain the output. For example, the weight of a truck on a bridge stresses the bridge, generating strain in the bridge structures. Specifically, stress is a measure of the force applied over a unit area and has units of pressure (N/mm^2). *Strain* is a relative deformation corresponding to the applied stress and has units of change in length divided by gauge length (stated in mm/mm or sometimes %). Detailed discussions of stress analysis are beyond the scope of this text. However, there are many mechanics books available for the interested reader for further information (for example, Oezkaya and Nordin, 1991; or Shigley and Mischke, 1996).

In the 1930s, the psychologist Hans Selye introduced the concept of stress (or distress, if excessive) as caused by *stressors*. He was borrowing an engineering term to describe a psychological condition. Yet the use of the term *stress* as either cause or result can create much confusion. (What is, for example, job stress?) To avoid ambiguity, the engineering terminology will be used in this text: stress produces strain.

MECHANICAL BASES

Mechanics is the study of forces and of their effects on masses. *Statics* considers masses at rest or in equilibrium as a result of balanced forces acting on them. In *dynamics,* one studies the motions of masses that are caused by unbalanced external forces.

Dynamics, often called kinesiology when applied to the human body, is again subdivided into two fields. *Kinematics* considers the motions (displacements and their time derivatives—velocity, acceleration, and jerk) but not the forces that bring these about; in contrast, *kinetics* is concerned with exactly these forces.

Newton's laws are basic to biomechanics. The first law states that a mass remains at uniform motion (which includes being at rest) until acted on by unbalanced external forces. The second law, derived from the first, indicates that force is proportional to the acceleration of a mass. The third law states that action is opposed by reaction.

Newton's second law sheds light on one important factor in biomechanics—force. Force is defined by its ability to change the velocity of a mass by generating acceleration: a force can cause a mass to speed up or slow down or change its direction of motion. Force is not a basic unit but a derived one, calculated from the acceleration of a mass. (Acceleration is the first time derivative of velocity.) Relatedly, no device exists that measures force directly. All measuring devices for force (or torque) rely on other physical phenomena, which are then transformed and calibrated in units of force (or torque). The events used to assess force are usually either displacement (such as bending of a metal beam) or acceleration experienced by a mass. Similarly, stress cannot be measured directly but, instead, is inferred from its effect on a measured quantity, such as strain or displacement over a known gauge length.

The correct unit for force measurement is the newton (N). One newton of force accelerates 1 kilogram of mass by 1 meter per squared second: $1 \text{ N} = 1 \text{ kg}$

* 1 m/s^{-2}. On earth, 1 kg has a weight (weight is a force!) of 9.81 N. Many older tables of forces use the units kg or kg_f or kp, or lb or lb_f, which can be translated into newtons because the kilogram-force unit equals 9.81 N and the pound-force unit is approximately 4.45 N.

Torque (also called moment) is the product of force and its lever arm (distance) to the articulation about which it acts. By definition, the lever arm is at a right angle to the direction of the force. In kinesiology, the lever arm is often called the mechanical advantage. Torque is defined by its ability to change the angular velocity of a mass by generating an angular acceleration. Torque, as well as force, is a vector, which means that it must be described not only in magnitude but also by direction (line of application) and its point of application.

To make use of force and moment, we must find a way to relate them. The approach is to enforce Newton's laws. In general, if there is no motion, or no change in motion, then we say that the object is in static equilibrium: static because there is no net motion and equilibrium because the forces and moments are balanced. Given balance, the sum of the forces and the sum of the moments must be zero:

$$\Sigma \text{ forces} = 0 \qquad\qquad (5\text{-}1a)$$

$$\Sigma \text{ moments} = 0 \qquad\qquad (5\text{-}1b)$$

Example:

If a person stands quietly on a scale, the weight due to the mass of the person must equal the force with which the scale is pushing back on the person's feet. The weight (mass times acceleration, according to Newton's second law) equals the reaction force (Newton's third law). In this case, the acceleration is that due to gravity, or 9.81 m/s^2.

To analyze the forces and moments acting on an object, we generally examine the object as a *free body*. The free body may be the entire person standing on the scale or may be only the arm of the person; what is selected depends on the desired results. At the boundaries of the free body, we must account for all the forces and moments due to the body's interaction with its environment (including gravity), often with other bodies. In the example of the person on the scale, if the person in his or her entirety is the free body, then we must include the weight (force) at the feet that causes the reaction on the scale.

Static Equilibrium: Analysis of Arm Lifting

Figure 5-1 uses the example of elbow flexion to illustrate these relationships. The primary flexing muscle, the biceps brachii, exerts a force **M'** on the forearm. The biceps force **M'** has a lever arm m around the elbow joint. The lever arm m is measured along a perpendicular line from the muscle force **M'** at its point of application to the point about which the motion takes place. This generates a torque **T**:

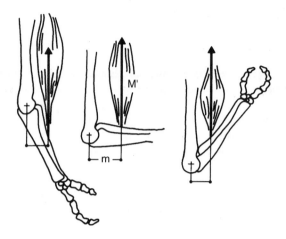

Figure 5-1.   Changing lever arm of the force vector **M**' with varying elbow angle.

$$T = m\mathbf{M}' \tag{5-2}$$

Since, by definition, there must be a right angle between lever arm and force vector, the lever arm m is smaller both when the arm is highly flexed or extended than when there is approximately a right angle between forearm and upper arm.

Figure 5-2 represents the same condition in a more realistic and detailed sketch. This shows that the actual direction of the muscle force vector **M** usually differs from that of the vector **P**, which is indeed perpendicular to the lever arm m. With the angle β between **P** and **M**, the relationship between the two vectors is

$$\mathbf{P} = \mathbf{M} \cos\beta \tag{5-3}$$

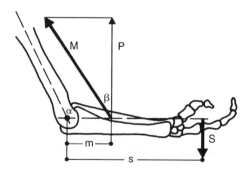

Figure 5-2.   Interaction between external force S and muscle force **M**.

The angle β itself varies with the elbow angle α (alpha); hence β is a function of α or, mathematically,

$$\beta = f(\alpha) \tag{5-4}$$

The torque $\mathbf{T}$ around the elbow joint generated by the muscle force $\mathbf{M}$ is

$$\mathbf{T} = m\mathbf{P} \tag{5-5}$$

If there is no acceleration, we know that all forces and torques must be balanced: in every direction, the sum of the forces and the sum of the torques must be zero. Thus, in this case, the torque $\mathbf{T}$ is counteracted by an equally large but opposing torque, which may be generated, for example, by a force vector $\mathbf{S}$ pulling down on the hand. With its lever arm s (assuming that the lever arm is perpendicular to the force), $\mathbf{S}$ establishes static equilibrium if

$$s\mathbf{S} = m\mathbf{P} = m\mathbf{M}\cos\beta \tag{5-6}$$

Measuring the external force $\mathbf{S}$ and knowing the lever arms s and m as well as the angle β, we can solve for the muscle force vector $\mathbf{M}$

$$\mathbf{M} = \left(\frac{\mathbf{S}}{\cos\beta}\right)\left(\frac{s}{m}\right) \tag{5-7}$$

Note that, in this example, the arm was considered massless. However, real-life conditions are a bit more complex. It is relatively easy to measure the lever arm s of the external force, but it can be rather difficult to determine the anatomic insertion point of the tendon connecting the biceps brachii muscle with the radius, which establishes the lever arm m of the muscle force vector $\mathbf{M}$. Furthermore, the muscle itself is partly restrained (by encapsulating ligaments) in a groove on the humerus (the bone of the upper arm) to near the elbow bend. This makes the angle β steeper near the elbow than it was assumed to be in the previous discussion and in Figure 5-2. (Also, the biceps muscle has two heads, which split along the upper part of the humerus and attach in different locations to the scapula, the shoulder blade.) In addition, other muscles help in flexing the elbow: the brachialis originates at about the half-length of the humerus on its anterior side and inserts on the ulna; and the brachioradialis can also contribute to elbow flexion.

Figure 5-3 illustrates these conditions. $\mathbf{M}$ is still the contractile force of the biceps, $(90°-\beta)$ is its pull angle with respect to the long axis of the forearm, and $\mathbf{P} = \mathbf{M}\cos\beta$ its torque-generating force about the elbow joint at the lever arm m. A similar vector analysis for the force vector $\mathbf{N}$ of the brachialis results in its torquing force $\mathbf{Q}$ at the lever arm n; angle γ (also a function of the elbow angle α) is the difference between the directions of $\mathbf{N}$ and $\mathbf{Q}$, hence $\mathbf{Q} = \mathbf{N}\cos\gamma$. Taking into account the external torque $\mathbf{T} = s\mathbf{S}$ as before (assuming a right angle between force $\mathbf{S}$ and its lever arm s; if this were not the case, a vector analysis of $\mathbf{S}$

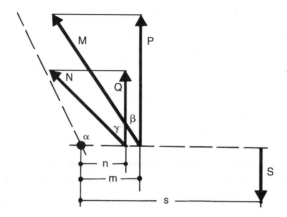

Figure 5-3. Interactions between several muscles and an external force. (Gravity disregarded.)

would have to be done to determine its angular components), the conditions for static equilibrium (i.e., no linear movements of the forearm and no change in elbow angle α; the arm is at rest) can be stated as follows.

There must be a horizontal joint reaction force **H** present so that the forearm will not be pushed away by the horizontal components of **M** (namely **M** $\sin\beta$) and of **N** (**N** $\sin\gamma$). The insertion of force **H** into equation 5-8 recognizes that need. **H** may be generated by shoulder muscles acting on the upper arm or by resting the elbow against a vertical stop.

All forces in the horizontal direction sum to zero:

$$\mathbf{M}\sin\beta + \mathbf{N}\sin\gamma + \mathbf{H} = 0 \qquad (5\text{-}8)$$

It is routine in this so-called free-body analysis to first insert joint reactions into the equations; experimental measurement, model assumptions, or mathematically solving the equations will make it clear whether these joint forces and torques actually exist or not.

Equation 5-9 states that the forearm will not be linearly elevated or lowered; for this, a joint reaction force **V** is inserted. All forces in the vertical direction sum to zero:

$$\mathbf{P} + \mathbf{Q} - \mathbf{S} + \mathbf{V} = 0 \qquad (5\text{-}9)$$

A joint-reaction torque **T** in Equation 5-10 assures that the elbow angle α does not change. All torques about the elbow joint sum to zero:

$$m\mathbf{P} + n\mathbf{Q} - s\mathbf{S} + \mathbf{T} = 0 \qquad (5\text{-}10)$$

If these three equations are satisfied, then the forearm is at rest with respect to the upper arm.

This brief example illustrates the importance of properly considering all the forces and torques that act upon and within the system. The segment hand-forearm was selected as the distal "free body" for the analysis. At the elbow joint, where the forearm connects with the upper arm, a reaction force **H** was introduced to provide horizontal restraint. Furthermore, joint reactions in vertical direction (force **V**) and torque **T** were inserted as requisites for static equilibrium. The next step in the free-body analysis is to proximally cross the free-body boundary at the elbow to consider the biomechanics of the upper arm in similar fashion. For this step in the analysis, the joint reaction forces and torques **H, V**, and **T** are transferred (in their opposite directions) to the distal end of the upper arm's free-body diagram. (For more information, check, for example, books by Chaffin and Andersson, 1991; Oezkaya and Nordin, 1991; or Shigley and Mischke, 1996.)

This discussion shows that it is relatively easy to measure the resultant output of all concurring muscular forces combined at a suitable interface between body limb and an external measuring device (here represented by force **S** at lever arm *s*), but the analysis does not indicate which muscles contribute how much. This is illustrated by the fact that the foregoing discussion did not take into account that the triceps muscle may also be involved for control and regulation; it would add to the torque generated by **S** and counteract the biceps and brachialis muscles, forcing them to increase their efforts. Hence, the analysis technique shows only "minimal net results," meaning that the calculated amounts of forces and moments are the minima needed to achieve the desired effect (in this case, no motion, or equilibrium). The actual efforts of individual muscles may be much greater than the equations indicate, especially if antagonistic groups of muscles counteract each other.

Dynamic Equilibrium: Gait Analysis

Biomechanics is especially useful for the analysis of the human body during activities of daily living. This becomes important in many dynamic applications, such as the evaluation of assisting devices for gait (i.e., canes, crutches, or walkers), efficacy of surgical procedures (i.e., total joint replacement with an artificial joint or limb lengthening), and usefulness of external protective equipment (i.e., back belts for use during lifting or knee braces for knee support and protection during sports).

The same approach as in the preceding arm example is taken for these analyses. First, an appropriate free body is chosen. For example, one might want to evaluate the forces on the knee joint during stance phase of walking with or without use of a cane. The free-body boundary can be chosen along the leg from the foot to the knee, cutting across the knee joint itself. Then the forces that cross the free-body boundary are identified, which must be accounted for in the free-body analysis. The foot-ground reaction force could be measured experimentally using a force plate; this force is expected to be different depending on whether the assisting device was used. Then there are forces at the free-body boundary at the knee: if the cut is made through the joint itself, we must consider the reaction force from the femur (thigh bone) to the tibia (shin bone) and the force acting through the patella (kneecap). Finally, using the geometry of the limb and the equilibrium conditions, the forces at the knee can be calculated and compared for the case of walking with or without a cane.

These biomechanical analyses can become very complex as more detailed models are constructed. We might consider only the major forces acting along bones and major muscle groups, or we might detail the contribution of each muscle, ligament, and other soft tissue around a joint, depending on the desired results. If the object to be analyzed is not at rest, then dynamic equilibrium considers the effects of changes in velocities (that is, accelerations) of body segments. This is often required for the analysis of running, jumping, cycling, or throwing since the motions are dynamic. For example, during a walking cycle, the leg moves with both linear and angular acceleration, that is, with changing velocities. Linear acceleration is along a straight, translatory line, while angular acceleration is in a rotation about an axis. To measure acceleration, accelerometers may be used; alternately, one could measure the position of the body segment over time and use numerical differentiation techniques to estimate the velocities and accelerations.

The equations for dynamic equilibrium become [see eqs. (5-1a) and (5-1b), for static equilibrium]:

$$\Sigma \text{ forces } = \text{mass} \times \text{acceleration} \tag{5-1a}$$

$$\Sigma \text{ moments } = (\text{mass moment of inertia}) \times (\text{angular acceleration}) \tag{5-1b}$$

Although it may become complicated to determine the mass properties and accelerations for the body segment under analysis, this has been the classical way to estimate joint and muscle forces (see e.g., Paul, 1967; Morrison, 1970; Ladin and Wu, 1991; Oezkaya and Nordin, 1991). One approach is simply to add dynamic components to the static terms in the equations, such as centrifugal, tangential, and Coriolis forces in rotations; other mathematical approaches are fundamentally different from those used in statics. In all cases, the basic mechanical principles remain the same: Newton's laws must be obeyed.

Recently, there have been successful attempts to measure the joint loads in the human body directly. For example, an artificial, special total hip replacement with a three-axes load cell in the neck of the metal component has been implanted in the thigh bone. The forces occurring during various activities have been telemetrically sent to recorders in the laboratory. The actual in vivo loads were found to be somewhat smaller than calculated previously (Davy et al., 1988; Bergmann, Graichen, and Rohlmann, 1993).

ANTHROPOMETRIC INPUTS

Biomechanics rely much on anthropometric data, adapted and often simplified to fit the mechanical approach. In Chapter 1, the reference planes used in anthropometry and, subsequently, in biomechanics were identified in Figure 1-1: the medial (midsagittal) plane divides the body into the right and left halves; the frontal (coronal) plane establishes anterior and posterior sections of the body; the transverse (horizontal) plane cuts the body into superior and inferior parts. However, anatomically, only the medial plane is well defined in its location, while the frontal and transverse planes need to be fixed by consensus. For this, it is usually assumed that the human stands upright in the so-called anatomical position and that, in this case, the three planes meet in the center of mass of the body (in the pelvic region), there establishing the origin of an *XYZ*-axis system. Obviously, this convention applies only to the upright standing body. If the posture is different, the location of the center of mass changes (see NASA/Webb, 1978; Roebuck, Kroemer, and Thomson, 1975; Hay, 1973). In this case, one may decide either to retain the anatomical fixation of the coordinate system in the pelvic area or to establish a new origin, depending on the given conditions.

Links and Joints

For ease of treatment and computation, the human skeletal system is often simplified into a relatively small number of straight-line links (representing long bones) and joints (representing major articulations). Figure 5-4 shows such a typical link-joint system. In this example, hands and feet are not subdivided into their components, and the spinal column is represented by only three links. Clearly, such simplification does not represent the true design of the human body but may be sufficient to represent certain mechanical properties, depending on the desired application.

The determination of the location of a joint's center of rotation is relatively easy for simple articulations, such as the hinge-type joints in fingers, elbows, and knees. However, this is much more difficult for complex joints with several degrees of freedom, such as in the hip, and even more difficult in the shoulder. In fact, the shoulder joint is depicted in Figure 5-4 as two joints with an intermediate scapular link. In many cases, the biomechanicist will simply but carefully assume location and properties of a joint in question so that it fits the given model requirements; hence, the model joint characteristics may reflect the true body articulation only partially or only within a limited range of motion

Once the joints are established, a straight-line link length is defined as the distance between adjacent joint centers. Tables 5-1 and 5-2 list the definitions of the joint centers and of the links between them.

Unfortunately, anthropometric dimensions are measured not between joints but usually between externally discernible landmarks, such as bony protrusions on the skeletal system. Hence, a major problem in developing a model depicting the human body is to establish the numerical relationships between standard anthropometric measures and link lengths. This usually cannot be done by simply expressing segment lengths as proportions of body stature, a procedure occasionally, and then usually falsely, propagated. As the correlation tables in Chapter 1 reflect, stature is highly related to some other height measures but shows only very low correlation coefficients with most other measures of the human body.

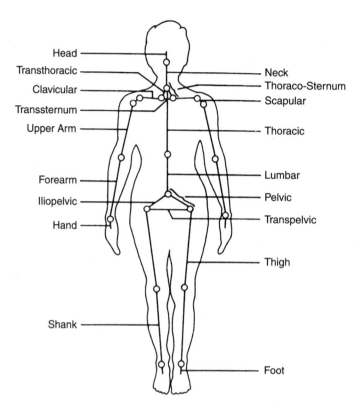

Figure 5-4. Typical link-joint system (NASA/Webb, 1978).

Hence, in many cases, one must derive link lengths from measures of bone lengths, which are not regularly taken in anthropometric surveys. However, if such data are available, many link lengths can be obtained using ratios or regression equations (see Tables 5-3 and 5-4). Other means to estimate length links, and the location of joint centers in relation to surface landmarks, are discussed by Chaffin and Andersson, 1991 and by NASA/Webb, 1978. Absolute measures of link lengths have been reported by NASA/Webb and by Gordon et al. (1989) and are listed in Table 5-5. When applying this information by adding or subtracting lengths, one must watch out for the effects of covariation; this is discussed in Chapter 1, where appropriate procedures are explained and formulas presented.

Body Volumes

Measurement of whole-body or of segment volume is often necessary to calculate inertial properties or to design close-fitting garments. Use of the Archimedes principle provides the body volume: The object is immersed in a container filled

Table 5-1.   Definitions of joint centers.

HEAD

Head/Neck	Midpoint of the interspace between the occipital condyle and the first cervical vertebra.
Neck/Thorax	Midpoint of the interspace between the 7th cervical and 1st thoracic vertebral bodies.

TRUNK

Thorax/Lumbar	Midpoint of the interspace between the 12th thoracic and 1st lumbar vertebral bodies.
Lumbar/Sacral	Midpoint of the interspace between the 5th lumbar and 1st sacral vertebral bodies.

LEG

Hip	(Lateral aspect) A point at the tip of the femoral trochanter 1.0 cm anterior to the most laterally projecting part of the femoral trochanter.
Knee	Midpoint of a line between the centers of the posterior convexities of the femoral condyles.
Ankle	Level of a line between the tip of the lateral malleolus of the fibula and a point 5 mm distal to the tibial malleolus.

ARM

Sternoclavicular	Midpoint position of the palpable junction between the proximal end of the clavicle and the sternum at the upper border (jugular notch) of the sternum.
Claviscapular	Midpoint of a line between the coracoid tuberosity of the clavicle (at the posterior border of the bone) and the acromioclavicular articulation (or the tubercle at the lateral end of the clavicle); the point should be visualized as on the underside of the clavicle.
Glenohumeral (shoulders)	Midregion of the palpable bony mass of the head and tuberosities of the humerus; with the arm abducted about 45 deg relative to the vertebral margin of the scapula, a line dropped perpendicular to the long axis of the arm from the outermost margin of the acromion approximately bisects the joint.
Elbow	Midpoint on a line between (1) the lowest palpable point, the medial epicondyle of the humerus, and (2) a point 8mm above the radiale (radiohumeral junction).
Wrist	On the palmar side of the hand, the distal wrist crease at the palmaris longus tendon, or the midpoint of a line between the radial styloid and the center of the pisiform bone; on the dorsal side of the hand, the palpable groove between the lunate and capitate bones, on a line with metacarpal bone III.

Source: Adapted from NASA/Webb 1978.

Table 5-2.   Definitions of links.

HEAD	The straight line between the occipital condyle/C1 interspace center and the center of mass of the head.
NECK	The straight line between the occipital condyle/C1 and the C7/T1 vertebral interspace joint centers.
TORSO (total)	The straight-line distance from the occipital condyle/C1 interspace joint center to the midpoint of a line passing through the right and left hip joint centers.

THORAX Sublink	Thoracosternum—A closed linkage system composed of three links. The right and left transthorax are straight-line distances from the C7/T1 interspace to the right and left sternoclavicular joint centers. The transternum link is a straight-line distance between the right and left sternoclavicular joint centers.
	    Clavicular—The straight line between the sternoclavicular and the claviscapular joint centers.
	    Scapular—The straight line between the claviscapular and glenohumeral joint centers.
	    Thoracic—The straight line between C7/T1 and T12/L1 vertebral body interspace joint centers.

LUMBAR Sublink	The straight line between the T12/L1 and L5/S1 vertebrae interspace joint centers.
PELVIS Sublink	Treated as a system composed of three links: the right and left iliopelvic links are straight lines between the L5/S1 interspace joint center and a hip joint center. The transpelvic link is a straight line between the right and left hip joint centers.
THIGH	The straight line between the hip and knee joint centers of rotation.
SHANK	The straight line between the knee and ankle joint centers of rotation.
FOOT	The straight line between the ankle joint center and the center of mass of the foot.
UPPER ARM	The straight line between the glenohumeral and elbow joint centers of rotation.
FOREARM	The straight line between the elbow and wrist joint centers of rotation.
HAND	The straight line between the wrist joint center of rotation and the center of mass of the hand.

Source: Adapted from NASA/Webb 1978.

Table 5-3. Ratios (in %) of link length to bone length.

Segment	Mean %	Standard deviation, %
Thigh link/Femur length	90.3	0.9
Shank link/Tibia length	107.8	1.8
Upper arm link/Humerus length	89.4	1.6
Forearm link/Ulna length	98.7	2.7
/Radius length	107.1	3.5

Source: Adapted from NASA/Webb, 1978, who refer to Dempster, 1964.

Table 5-4. Regression equations for estimating link lengths (in cm) directly from bone lengths.

Empirical equation		Standard error of estimate	Correlation coefficient
Forearm link length	= 1.0709 radius length	NA	NA
	= 0.9870 ulna length	NA	NA
Arm link length	= 66.2621 + 0.8665 ulna length	9.90	0.94
	= 58.0752 + 0.9646 radius length	8.92	0.94
Thigh link length	= 132.8253 + 0.8172 tibia length	16.57	0.73
	= 92.0397 + 0.8699 fibula length	10.34	0.87
Shank link length	= 8.2184 + 1.0904 fibula length	5.95	0.97
	= 1.0776 tibia length	NA	NA

Source: Adapted from NASA/Webb 1978.

Table 5-5. Estimated link lengths (in cm) for 1985 U.S. population.

Limb	*Male*			*Female*		
	5th	50th	95th	5th	50th	95th
Upper arm link	28.6	30.5	32.3	26.1	27.8	29.5
Forearm link (ulna)	25.6	27.1	28.7	22.7	24.1	25.5
Forearm link (radius)	25.9	27.5	29.2	22.7	24.1	25.5
Thigh link	40.4	43.2	46.1	36.9	39.5	42.1
Shank link	38.9	42.1	45.3	34.7	37.4	40.0

Source: Adapted from NASA/Webb, 1978.

with water; measuring the volume of the displaced water yields the volume of the immersed body. The technique works well for obtaining volumes of limbs and also of the whole body if changes due to respiration can be controlled.

There are also indirect methods to obtain the volume. One is to assume that body segments can be represented by geometrically known forms—such as sphere, cylinder, or truncated cone—for which the volume can be calculated easily; or one may use information about the cross-section contours, which can be obtained by CAT scans or by dissection. If these cross-section contours are taken at sufficiently close separations so that the changes between cross sections can be assumed linear with distance, the volume V can be calculated by summing the cross-section areas A, multiplied with their distances d from each other:

$$V = \Sigma(A_i d_i) \qquad (5\text{-}11)$$

Often, the distance d between cross-section costs is kept constant, and adjacent cross-sectional areas are averaged:

$$\tilde{A}_i = \tfrac{1}{2}(A_{i-1} + A_i) \qquad (5\text{-}12)$$

Other approximations rely on the assumption that body segments resemble regular geometric figures. For example, if the body cross section is elliptical, then its area can be calculated by

$$A_i = \pi\, a_i\, b_i \qquad (5\text{-}13)$$

where a_i is the semimajor axis and b_i is the semiminor axis of section i. The volume can then be calculated according to equation (5-11). Of course, one may assume that body segments resemble cylindrical cone sections or that they can be represented by cylinders. Such simplifying assumptions lend themselves to easier, but possibly less accurate, calculations of the volume.

Inertial Properties

Knowledge of the total body mass and its distribution throughout the body is important for the assessment of static and dynamic properties of the human body. To obtain such data, many methods and techniques have been developed. The publications by Clauser, McConville, and Young (1969), Kaleps et al. (1984), McConville et al. (1980) provide historically interesting reviews and complete data sets; also, see Chaffin and Andersson (1984), Hay (1973), NASA/Webb (1978), Roebuck, Kroemer, and Thomson (1975), and Roebuck (1995) for data compilations and discussions.

The simplest inertial property is weight, a force that can be measured easily with a variety of scales. Properly, the inertial property should be the mass, which is the weight divided by the acceleration due to gravity. However, it is common to measure the weight of objects on earth. (Human body weight is usually measured in air; hence, there is a slight error due to buoyancy.) Using cadaver data, body segment weight can be predicted from total body weight (see Table 5-6).

According to Newton's second law, weight **W** is a force depending on body mass m and the gravitational acceleration g, according to

$$\mathbf{W} = mg \tag{5-14}$$

Density and Mass

Density D is the mass per unit volume, where V is the volume of the object in question:

$$D = \frac{W}{gV} \tag{5-15}$$

Mass is, of course,

$$m = DV \tag{5-16}$$

Table 5-6. Prediction equations to estimate segment mass (in kg) from total body weight **W** (in kg).

Segment	Empirical equation	Standard error of estimate	Correlation coefficient
Head	0.0306 W + 2.46	± 0.43	0.626
Head and neck	0.0534 W + 2.33	± 0.60	0.726
Neck	0.0146 W + 0.60	± 0.21	0.666
Head, neck, and torso	0.5940 W − 2.20	± 2.01	0.949
Neck and torso	0.5582 W − 4.26	± 1.72	0.958
Total arm	0.0505 W + 0.01	± 0.35	0.829
Upper arm	0.0274 W − 0.01	± 0.19	0.826
Forearm and hand	0.0233 W − 0.01	± 0.20	0.762
Forearm	0.0189 W − 0.16	± 0.15	0.783
Hand	0.0055 W + 0.07	± 0.07	0.605
Total leg	0.1582 W + 0.05	± 1.02	0.847
Thigh	0.1159 W − 1.02	± 0.71	0.859
Shank and foot	0.0452 W + 0.82	± 0.41	0.750
Shank	0.0375 W + 0.38	± 0.33	0.763
Foot	0.0069 W + 0.47	± 0.11	0.552

Source: Adapted from NASA/Webb 1978.

The specific density D_{sp} is the ratio of D to the density of water, D_w

$$D_{sp} = \frac{D}{D_w} \qquad (5\text{-}17)$$

The human body is not homogeneous throughout; its density varies depending on cavities, water content, fat tissue, bone components, and so forth. Still, in many cases, it is sufficient to assume that either the body segment in question or even the whole body is of constant (average) density.

Information about the relations of body segment weight to total body weight, compiled in Table 5-7, provides some insight into mass and density distributions. Detailed density data have been obtained from weight and volume measurements of dissected cadavers (see Tables 5-8 and 5-9). However, there are obvious difficulties associated with the use of cadaver material, such as loss of fluids, chemical changes in tissue, and preservation and biohazard considerations. Hence, other studies have used immersion and weighing techniques to determine density values on living subjects. Stereophotometry has been combined with classic anthropometric techniques to establish mass properties of living persons (Kaleps et al., 1984).

Another useful concept to distinguish between the compositions of different bodies is that of the lean body mass or lean body weight. This relies on the assumption that basic structural body components such as skin, muscle, bone, and so forth, are relatively constant in percentage composition from individual to individual.

Table 5-7. Relative weights of body segments (in %).

Groups of segments as % of total body weight	Single segments as % of segment groups
Head and neck = 8.4	Head = 73.8
	Neck = 26.2
Torso = 50.0	Thorax = 43.8
	Lumbar = 29.4
	Pelvis = 26.8
Total leg = 15.7	Thigh = 63.7
	Shank = 27.4
	Foot = 8.9
Total arm = 5.1	Upper arm = 54.9
	Forearm = 33.3
	Hand = 11.8

Source: Adapted from NASA/Webb, 1978.

Table 5-8. Segmental mass ratios, in percent, derived from cadaver studies (adapted from Roebuck, Kroemer, and Thomson 1975).

	Harless (1860)	Braune and Fischer (1889)	Fischer (1906)	Dempster* (1955)	Clauser, McConville and Young (1969)	Average
Sample Size	2	3	1	8	13	
Head	7.6	7.0	8.8	8.1	7.3	7.8
Trunk	44.2	46.1	45.2	49.7	50.7	47.2
Total arm	5.7	6.2	5.4	5.0	4.9	5.4
Upper arm	3.2	3.3	2.8	2.8	2.6	2.9
Forearm and hand	2.6	2.9	2.6	2.2	2.3	2.5
Forearm	1.7	2.1	—	1.6	1.6	1.8
Hand	0.9	0.8	—	0.6	0.7	0.8
Total leg	18.4	17.2	17.6	16.1	16.1	17.1
Thigh	11.9	10.7	11.0	9.9	10.3	10.8
Shank and foot	6.6	6.5	6.6	6.1	5.8	6.3
Shank	4.6	4.8	4.5	4.6	4.3	4.6
Foot	2.0	1.7	2.1	1.4	1.5	1.7
Total body**	100.0	100.0	100.0	100.0	100.0	100.0

*Dempster's values adjusted by Clauser, McConville, and Young (1969).
**Calculated from head + trunk + 2 (total arm + total leg).

However, the component fat varies in percentage of total mass or weight throughout the body and for different persons. This allows us to express body weight W as

$$W = \text{lean body weight} + \text{fat weight} \tag{5-18}$$

There are several techniques to determine body fat. Many use skinfold measures whereby, in selected areas of the body, the fold thickness of loose skin is measured with special calipers. Unfortunately, measurement of skinfold thickness is a rather difficult and not very reliable procedure since skinfolds may be grasped and compressed differently, they may slip from the instrument, and the pressure applied by the instrument over its measuring surfaces may vary.

Locating the Center of Mass

The body mass may be considered as concentrated at one point in the body where its physical characteristics respond in the same way as distributed throughout the

Table 5-9. Densities of the total body of cadavers and their segments.

	Harless (1860)	Dempster (1955)		Clauser et al. (1969)	
	Mean	**Mean**	**SD**	**Mean**	**SD**
Total body	—	—	—	1.042	.018
Head	—	—	—	1.071	—
Neck and torso	—	—	—	1.023	.032
Head and neck	1.11	1.11	.012	—	—
Thorax	—	0.92	.056	—	—
Abdomino-pelvic	—	1.01	.014	—	—
Thigh	1.07	1.05	.008	1.045	.017
Shank	—	1.09	.015	1.085	.014
Foot	1.09	1.10	.056	1.085	.014
Arm	—	1.07	.027	1.058	.025
Forearm	1.10	1.13	.037	1.099	.018
Hand	1.11	1.16	.110	1.108	.019
Torso-limbs	—	1.07	.016	—	—
Upper arm	1.08	—	—	—	—
Lower leg	1.10	—	—	—	—

Source: Adapted from McConville, et al., 1980.

body. This becomes particularly important for analyses of gait and other motions. The measurement of the location of the mass center is somewhat difficult with living persons because respiration causes shifts in the mass distribution, as do muscular contractions and food and fluid ingestion or excretion. Of course, there are major shifts in the location in the center of mass with changed body positions and, in particular, with body movements.

For the body at rest, various methods exist to determine the location of the center of mass CM. Most rely on the principle of finding the one location at which a single support would keep the body balanced. One of the simplest techniques is to place the body on a platform supported by two scales at precisely known support points. The body weight is then counteracted by the two forces at the support points and, with their distances from the line of action of the body weight, the balancing moments can be calculated. From this, their lever arms (distances) can be determined. Figure 5-5 shows this procedure, and the calculations in the plane shown proceed as follows.

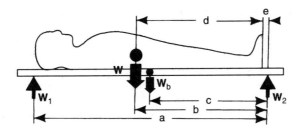

Figure 5-5. Finding the center of mass of the body placed on a known board and on two scales.

The sum of the vertical components of force and the sum of the moments about $\mathbf{W}_2$ are both zero, since the body is at rest:

$$\mathbf{W}_1 - \mathbf{W} - \mathbf{W}_b + \mathbf{W}_2 = 0$$

$$a\mathbf{W}_1 - b\mathbf{W} - c\mathbf{W}_b = 0 \qquad (5\text{-}19)$$

and, geometrically,

$$b = d + e \qquad (5\text{-}20)$$

where $\mathbf{W}_1$ is the force at scale 1, a the distance between scales 1 and 2, $\mathbf{W}$ the weight of the body at its center of mass CM, $\mathbf{W}_b$ the weight of the board at its CM, and c its distance from $\mathbf{W}_2$; d is the distance between the CM of the body and the soles of the feet, and e is their distance from $\mathbf{W}_2$.

Rearranging and inserting known force and distance values provide the solution:

$$d = \frac{a\mathbf{W}_1 - c\mathbf{W}_b}{\mathbf{W}_1 + \mathbf{W}_2 - \mathbf{W}_b} - e \qquad (5\text{-}21)$$

Note that the actual weight $\mathbf{W}$ of the subject need not be known.

This method can be used, of course, with the subject in any body position; hence, the distances in the X and Z directions can be determined for the center of mass similarly to the determination in the Y direction.

Body Segment Mass

Table 5-7 reflects ratios between segment and total body masses, and Table 5-10 lists relative locations of mass centers; these data stem from several studies, some of which are more than a century old. For further information about mass properties

Table 5-10. Locations of the centers of mass of body segments, measured in percent from their proximal ends.

	Harless (1860)	Braune and Fischer (1889)	Fischer (1906)	Dempster (1955)	Clauser, McConville and Young (1969)
Sample Size	2	3	1	8	13
Head (from top)	36.2	—	—	43.3	46.6
Trunk*	44.8	—	—	—	38.0
Total arm	—	—	44.6	—	41.3
Upper arm	—	47.0	45.0	43.6	51.3
Forearm and hand*	—	47.2	46.2	67.7	62.6
Forearm*	42.0	42.1	—	43.0	39.0
Hand*	39.7	—	—	49.4	18.0
Total leg*	—	—	41.2	43.3	38.2
Thigh*	48.9	44.0	43.6	43.3	37.2
Shank and foot	—	52.4	53.7	43.7	47.5
Shank	43.3	42.0	43.3	43.3	37.1
Foot	44.4	44.4	—	42.9	44.9
Total body (from head)	41.4	—	—	—	41.2

*The values on these lines are not directly comparable since the different investigators used differing definitions for segment lengths.

Source: Adapted from Roebuck, Kroemer, and Thomson, 1975.

of the human body and their assessment, see, for example, publications by Chaffin and Andersson (1991), Chandler et al. (1975), Hay (1973), Kaleps et al. (1984), McConville et al. (1980), NASA/Webb (1978), Roebuck, Kroemer, and Thomson (1975), and Roebuck (1995).

For the modeling of rotation movements, one needs information about the moments of inertia. There are three common approaches to determine the principal moments of inertia of the body and its segments. The first is to use geometric models, assuming that the selected shapes (spheres, cylinders, etc.) represent the actual body form and that their densities are constant and known. The second technique uses the radius of gyration K, expressed as a portion of segment (or link) length L. This length L is multiplied by K, such as listed in Table 5-11. The resulting product is multiplied by the appropriate segment mass to obtain an estimate of the moment of inertia.

The third approach to predicting the principal moment of inertia is to use regression equations based on body weight, segment weight, or segment volume. Few studies have been conducted to ascertain this information; among them, the report on men by McConville et al. (1980) and the one on women by Young et al. (1983) stand out. They photographed 31 males and 46 women, respectively and,

Table 5-11. Radius of gyration K in percent of segment length, L. The x, y, and z are the principal axes as shown in Figures 1-1 and 5-9.

Name		Link L	K, in %
Head	x	Head length	31.6
	y		30.9
	z		34.2
Torso	x	Torso length	43.0
	y	(Suprasternale height	35.2
	z	to trochanterion height.)	20.8
Thigh	x	Trochanterion height	27.9
	y	to fibular height	28.4
	z		12.2
Shank	x	Fibular height	28.2
	y		28.2
	z		7.6
Foot	x	Foot length	26.1
	y		24.9
	z		12.2
Upper arm	x	Acromion to radiale length	26.1
	y		25.4
	z		10.4
Forearm	x	Radiale to stylion length	29.6
	y		29.2
	z		10.8
Hand	x	Hand breadth	50.4
	y		45.6
	z		26.6

Source: Adapted from NASA/Webb, 1978.

from these records, determined stereometrically body segment volumes, moments of inertia, and their principal axes related to anatomical landmarks. The reports contain multiple regression equations to predict mass distribution properties from anthropometric measures.

KINEMATIC CHAIN MODELS

The stick person concept, which utilizes links and joints, embellished with volumes and masses and driven by muscles, can often be used to model human motion and strength capabilities.

Figure 5-6 shows a model of the human body related to the one shown in Figure 5-4. Forces exerted with the hand to an outside object ($\mathbf{H}_x$, $\mathbf{H}_y$, or $\mathbf{H}_z$) or

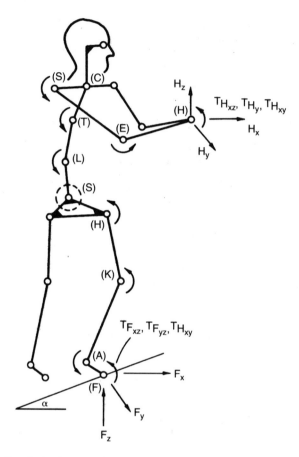

Figure 5-6. Free-body diagram of the link-joint model of the human body, indicating the chain of forces (F) or torques (T) transmitted from the hand through arm, trunk, and legs to the foot on the floor.

the torques generated in the hand ($\mathbf{T}_H$) are transmitted along the links. First, the force exerted with the right hand, modified by the existing mechanical advantages, must be transmitted across the right elbow (E). (Also, at the elbow, additional force or torque must be generated to support the mass of the forearm. However, for the moment, the model will be considered massless.) Similarly, the shoulder (S) must transmit the same efforts, again modified by existing mechanical conditions. In this manner, all subsequent joints transmit the effort exerted with the hands throughout the trunk, hips, and legs and, finally, from the foot to the floor. Here, the orthogonal force and torque vectors can be separated again, similarly to the vector analysis at the hand. Still assuming a massless body, the same sum of vectors must exist at the feet as was found at the hand.

Of course, the assumption of zero body mass is unrealistic. This can be reme-
died by incorporating information about mass properties of the human body segments, as mentioned earlier. Furthermore, considering only the efforts visible at
the interfaces between the body and the environment disregards the fact that, at
all body joints, antagonistic muscle groups exist that may counteract each other
for control and stabilization. Their possibly high individual efforts may nullify
each other to the outside observer.

Also, this example does not consider three-dimensional conditions. Finally,
body motions (accelerations), instead of the static position assumed here, would
further complicate the model.

However, if this simple model is accepted, a set of equations allows its com-
putational analysis. These equations follow from the standard procedure of setting the sum of forces in all directions to zero and likewise the sums of all
torques, for this body at rest (in equilibrium).

$$\mathbf{H}_x + \mathbf{F}_x = 0 \qquad\qquad (5\text{-}22)$$

$$\mathbf{H}_y + \mathbf{F}_y = 0 \qquad\qquad (5\text{-}23)$$

$$\mathbf{H}_z + \mathbf{F}_z = 0 \qquad\qquad (5\text{-}24)$$

$$\mathbf{T}_{\mathbf{H}xz} + \mathbf{T}_{\mathbf{F}x} = 0 \qquad\qquad (5\text{-}25)$$

$$\mathbf{T}_{\mathbf{H}y} + \mathbf{T}_{\mathbf{F}yz} = 0 \qquad\qquad (5\text{-}26)$$

$$\mathbf{T}_{\mathbf{H}x} + \mathbf{T}_{\mathbf{F}xy} = 0 \qquad\qquad (5\text{-}27)$$

In this example, it is assumed that one foot standing on a plane inclined by
the angle α counteracts all the hand efforts.

Following such procedures, rather complex models of the human body can
be developed, as shown in detail by Chaffin and Andersson (1991).

DESCRIPTION OF HUMAN MOTION

Classically, human movement has been described with anatomically derived
terms used in the medical profession. Unfortunately, there are systematic and
deep-rooted problems associated with these terms: motions are considered to
begin from the so-called anatomical position as depicted earlier in Figure 1-1, but
with the palms of the hands twisted forward, that is, supinated. Words to describe
motions were generated using the stems *flexion, tension, duction, rotation,* and
nation, together with the prefixes *ex-, hyper-, ad-, ab-, inward-, outward-, supi-,*
and *pro-* (see Fig. 5-7). Unfortunately, the same terms are applied indiscriminately to rotational movements about joints and to translational displacements of

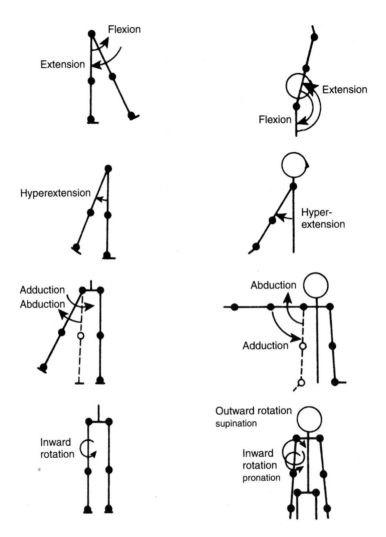

Figure 5-7.   Classic notation of body motions.

limbs; certain motions said to occur in a given plane may also occur in others. For example, rotation of the arm can occur in the sagittal, frontal, or transverse planes and in practically any other orientation. Severe difficulties are associated with motion in the shoulder joint: having "abducted" the upper arm 90 deg sideways, the limb actually moves toward the body centerline ("adduction") in whatever motion is performed. Similar problems arise when we try to describe inward and outward rotation of the forearm. It is very difficult to describe relative locations of body parts, and a "zero position" is often not specified (see Chap. 2).

In 1968, Roebuck made a thorough and radically new attempt to generate a more appropriate terminology (described in detail by Roebuck, Kroemer, and Thomson, 1975), using a coordinate system that is attached to the human pelvic area, as shown in Figure 1-1. The common frontal, sagittal, and transverse planes are used, and the usual 360-deg rotation is centered at each body articulation considered. The important basic concept is to describe consistently movements of the limbs with respect to their joints, as referred to the overall coordinate reference point in the pelvic area. Roebuck's system uses the notation of a clockwise/counterclockwise (as seen by the subject) rotation and of "zero" being footward in the frontal and sagittal planes, and forward in the transverse plane (see Figs. 5-8 and 5-9).

To describe movement verbally, Roebuck used the word *vection* (from the Latin *vehere,* "to carry") which, unfortunately, did not find widespread acceptance in spite of the fact that it replaced both *duction* and *nation.* Hence, it was proposed that the common *twist* be used to indicate rotation about a long axis (*nation*), while the familiar *flexion* and *extention* be maintained, with *pivot* indicating rotation about an axis perpendicular to the flexion-extension axis (Kroemer et al., 1989). Thus, for example, the index finger may be flexed or extended, or pivoted ulnarly or radially. Roebuck's prefixes are kept, *e-* signifying out or up from the standard position; *in-* the opposite of *e-,* that is, in or down; *posi-* and *negi-* are

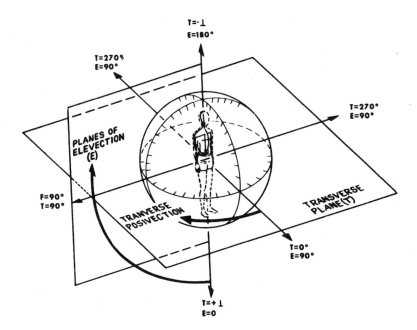

Figure 5-8. Roebuck's global coordinate system (adapted from Roebuck, Kroemer, and Thomson, 1975).

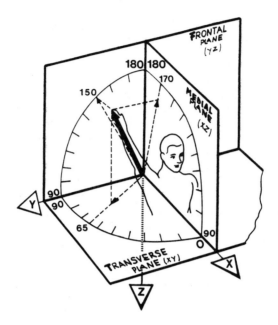

Figure 5-9. Description of the angular position of the right arm in the new notation system (adapted from Roebuck, Kroemer, and Thomson, 1975). The position of the hand point on the arm vector is $F = 150$ deg, $S = 80$ deg, $T = 25$ deg. Note that the directions for x, y, and z are commonly used in biomechanics (NASA, 1989) to describe accelerations of the vehicle that carries a human occupant.

self-explanatory. Table 5-12 lists the terms. Any motion can be described in this terminology by a descriptive term composed of the designations for the plane, the direction, and the type of motion. For example, *sagflexion* indicates that the motion occurs in the sagittal plane.

Consistent use of reference points, planes, and motion terms should reduce uncertainty about human motion capabilities and allow its measurement and reporting utilizing computer models and systems. For data regarding current information about motion capabilities, see Chapter 2. For instruments suitable for measuring joint mobility, check Chaffin and Andersson (1991), Roebuck, Kroemer, and Thomson (1975), Roebuck (1995), or Radwin et al. (1996).

SUMMARY

Biomechanical modeling of the human body or of its parts normally requires simplification of actual (physiologic, anatomic, anthropometric, etc.) characteristics to fit the methods and techniques derived from mechanics. While this establishes limitations regarding the completeness, reliability, and validity of the biomechanical procedures, it also allows research and conclusions with unique insights that

Table 5-12.   New terminology to describe body motions.

New terms	Meaning	Replacing
Pivot and flexion, extension	Rotation of a body segment about its proximal joint	duction vection rotation
Twist	rotation of a body segment about its internal axis	-nation rotation
Ex-	away (up, out) from zero	ab- e-
In-	toward (in, down) zero	ad-
Clock(wise) (or none)	in clockwise direction, as seen on own body	supi- or pro- (depending on body segment)
Counter(clockwise)	in counterclockwise direction, as seen on own body	pro- or supi- (depending on body segment)
Front-, trans-, sag-	reference to the plane in which motion is described	uncertainty and confusion

Examples:

Front-ex-pivot = pivoting movement in the frontal plane, away from zero
Trans-in-twist = twisting movement in the transverse plane, clockwise

would not have been possible using traditional (physiological, anatomical, or anthropometric) approaches. However, we have to be keenly aware of the limitations imposed by the underlying simplifications.

An important application of biomechanics is in the calculation of torques or forces that can be developed by muscles about body joints. In reversing this procedure, we can assess the strain on muscles, bones, and tissues generated by external loads on the body in various positions.

Body segment dimensions, their volumes, and mass properties can be calculated from anthropometric data.

Kinematic chain models of linked body segments allow the prediction of total-body capability (e.g., lifting) from the consideration of body segment capabilities.

A taxonomy for position and motion of the body exists that describes these both mathematically and verbally.

REFERENCES

Braune, W., and Fischer, O. (1889). Determination of the Moments of Inertia of the Human Body and Its Limbs. In German; translation in (1963) *Human*

Mechanics. AMRL-TDR-63-123. Wright-Patterson AFB, OH: Aerospace Medical Research Laboratory.

Bergmann, G., Graichen, F., and Rohlmann, A. (1993). Hip Joint Loading During Walking and Running, Measured in Two Patients. *Journal of Biomechanics* 26(8):969–990.

Chaffin, D.B. and Andersson, G.B.J. (1984). *Occupational Biomechanics.* New York, NY: Wiley.

Chaffin, D.B. and Andersson, G.B.J. (1991). *Occupational Biomechanics,* 2nd ed. New York, NY: Wiley.

Chandler, R.F., Clauser, C.E., McConville, J.R., Reynolds, H.M., and Young, J.W. (1975). *Investigation of Inertial Properties of the Human Body.* AMRL-TR-74-137. Wright-Patterson AFB. OH: Aerospace Medical Research Laboratory.

Clauser, C.E., McConville, J.T., and Young, J.W. (1969). *Weight, Volume, and Center of Mass of Segments of the Human Body.* AMRL-TR-69-70. Wright-Patterson AFB, OH: Aerospace Medical Research Laboratory.

Davy, D.T., Kotzar, G.M., Brown, R.H., Heiple, K.G., Goldberg, V.M., Heiple, K.G. Jr., Bevilla, J., and Burstein, A.H. (1988). Telemetric Force Measurements across the Hip after Total Arthroplasty. *Journal of Bone and Joint Surgery* 70(A):45–50.

Gordon, C.C., Churchill, T., Clauser, C.E., Bradtmiller, B., McConville, J.T., Tebbetts, I., and Walker, R.A. (1989). *1988 Anthropometric Survey of U.S. Army Personnel: Summary Statistics Interim Report.* Natick-TR-89/027. Natick, MA: U.S. Army Natick Research, Development and Engineering Center.

Hay, J.G. (1973). The Center of Gravity of the Human Body. In *Kinesiology III.* Washington, DC: American Association for Health, Physical Education, and Recreation, pp. 20–44.

Kaleps, I., Clauser, C.E., Young, J.W., Chandler, R.F., Zehner, G.F., and McConville, J. (1984). Investigation into the Mass Distribution Properties of the Human Body and Its Segments. *Ergonomics* 27(12):1225–1237.

King, A.I. (1984). A Review of Biomechanical Models. *Journal of Biomechanical Engineering* 106:97–104.

King, A.I., and Chou, C.C. (1976). Mathematical Modeling, Simulation and Experimental Testing of Biomechanical System Crash Response. *Journal of Biomechanics* 9:301–317.

Kroemer, K.H.E., Marras, W.S., McGlothlin, J.D., McIntyre, D.R., and Nordin, M. (1989). *Assessing Human Dynamic Muscle Strength.* Technical Report, 8-30-89. Blacksburg, VA: Virginia Tech, Industrial Ergonomics Laboratory. Also published (1990) in *International Journal of Industrial Ergonomics* 6:199–210.

Kroemer, K.H.E., Snook, S.H., Meadows, S.K., and Deutsch, S. (eds.) (1988). *Ergonomic Models of Anthropometry, Human Biomechanics and Operator-Equipment Interfaces.* Washington, DC: National Academy of Sciences.

Ladin, Z., and Wu, G. (1991). Combining Position and Acceleration Measurements for Joint Force Estimation. *Journal of Biomechanics* 24(12): 1173–1187.

McConville, J.T., Churchill, T., Kaleps, I., Clauser, C.E., and Cuzzi, J. (1980). *Anthropometric Relationships of Body and Body Segment Moments of Inertia.*

AFAMRL-TR-80-119. Wright-Patterson AFB, OH: Aerospace Medical Research Laboratory.

NASA (1989). *Man-Systems Integration Standards, Revison A.* NASA-STD 3000. Houston, TX: L.B.J. Space Center, SP34-89-230.

NASA/Webb (1978). *Anthropometric Sourcebook* (3 vols.). NASA Reference Publication 1024. Houston, TX: L.B.J. Space Center, NASA (NTIS, Springfield, VA 22161, Order 79 11 734).

Morrison, J.B. (1970). The Mechanics of the Knee Joint in Relation to Normal Walking. *Journal of Biomechanics* 3:51–61.

Oezkaya, N., and Nordin, M. (1991). *Fundamentals of Biomechanics.* New York, NY: Van Nostrand Reinhold.

Paul, J.P. (1967). Forces Transmitted by Joints in the Human Body. *Proceedings of the Institute of Mechanical Engineers 1966–67* 181(3J):8–15.

Radwin, R.C., Beebe, D.J., Webster, J.G., and Yen, T.Y. (1996). Instrumentation for Occupational Ergonomics. Chapter 7 in A. Bhattacharia and J.D. McGlothlin (eds.). *Occupational Ergonomics. Theory and Applications.* New York, NY: Marcel Dekker, pp. 165–193.

Roebuck, J.A. (1995). *Anthropometric Methods. Designing to Fit the Human Body.* Santa Monica, CA: Human Factors and Ergonomics Society.

Roebuck, J.A., Kroemer, K.H.E., and Thomson, W.G. (1975). *Engineering Anthropometry Methods.* New York, NY: Wiley.

Shigley, J.E. and Mischke, C.R. (eds.). *Standard Handbook of Machine Design* (2nd ed.) New York, NY: McGraw-Hill.

Young, J.W., Chandler, R.F., Snow, C.C., Robinette, K.M., Zehner, G.F., and Lofberg, M.S. (1983). *Anthropometric and Mass Distribution Characteristics of the Adult Female.* FAA-AM-83-16. Oklahoma City, OK: FAA Civil Aeromedical Institute.

FURTHER READING

Bagchee, A., and Bhattacharya, A. (1996) Biomechanical aspects of body movement Chapter 4 in A. Bhattacharya and J.D. McGlothlin (eds.). *Occupational Ergonomics.* New York, NY: Marcel Dekker, pp. 77–113.

Beer, F.P., and Johnston, E.R. Jr. (1981). *Mechanics of Materials.* New York, NY: McGraw-Hill.

Burstein, A.H., and Wright, T.M. (1994). *Fundamentals of Orthopaedic Biomechanics.* Baltimore, MD: Williams & Wilkins.

Chaffin, D.B. and Andersson, G.B.J. (1991). *Occupational Biomechanics,* 3rd ed. New York, NY: Wiley.

Fung, Y.C. (1993). *Biomechanics: Mechanical Properties of Living Tissues,* 2nd ed. New York, NY: Springer-Verlag.

Kenedi, R.M. (ed) (1980). *A Textbook of Biomedical Engineering.* London, UK: Blackie & Son.

Kroemer, K.H.E., Snook, S.H., Meadows, S.K., and Deutsch, S. (eds.) (1988). *Ergonomic Models of Anthropometry, Human Biomechanics and Operator-Equipment Interfaces.* Washington, DC: National Academy of Sciences.

Mow, C., and Hayes, W.C. (eds.) (1991). *Basic Orthopedic Biomechanics.* New York, NY: Raven Press.

Oezkaya, N., and Nordin, M. (1991). *Fundamentals of Biomechanics.* New York, NY: Van Nostrand Reinhold.

Roebuck, J.A. (1995). *Anthropometric Methods. Designing to Fit the Human Body.* Santa Monica, CA: Human Factors and Ergonomics Society.

The following listing of journals may be useful to keep track of new information in biomechanics. Of course, other publications also carry related articles.

Applied Ergonomics
Clinical Biomechanics
International Journal of Industrial Ergonomics
Journal of Biomechanical Engineering
Journal of Biomechanics
Journal of Bone and Joint Surgery
Journal of Hand Surgery
Ergonomics
Human Factors
Kinesiology
Occupational Ergonomics
Orthopedics
Spine

Chapter 6

The Respiratory System

OVERVIEW

The respiratory system provides oxygen for the energy metabolism and dissipates metabolic by-products. In the lungs, oxygen is absorbed into the blood. The circulatory system transports oxygen (and nutrients) throughout the body, particularly to the working muscles. It removes metabolic by-products either to the skin (where water and heat are dissipated) or to the lungs. There, carbon dioxide, as well as water and heat, is dispelled while more oxygen is absorbed.

THE MODEL

Figure 6-1 indicates the close interaction between the respiratory system, which absorbs oxygen and dispels carbon dioxide, water, and heat, and the circulatory system, which provides the means of transport.

INTRODUCTION

The respiratory system moves air to and from the lungs, where part of the oxygen contained in the inhaled air is absorbed into the bloodstream; it also removes carbon dioxide, water, and heat into the air to be exhaled. The absorption of oxygen and the expulsion of metabolites takes place at spongelike surfaces in the lungs that contain small air sacs (alveoli).

ARCHITECTURE

The so-called respiratory tree in Figure 6-2 shows schematically the path of the air that has entered the body via the nose and mouth and passed through the throat

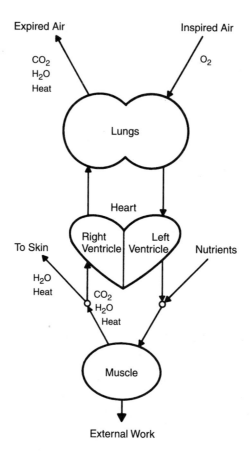

Figure 6-1.   Scheme of the interrelated functions of the respiratory and circulatory systems.

(pharynx), voice box (larynx), and windpipe (trachea) into the bronchial tree. Here, the airways branch repeatedly until they terminate in the alveoli of the lungs. These are small air sacs with diameters of about 400 μm, with thin membranes between capillaries and other subcellular structures allowing the exchange of oxygen (O_2) and carbon dioxide (CO_2) between blood and air.

The adult has 23 subsequent dichotomic branching steps. The first 16 branches merely conduct the air with little or no gas exchange. The next 3 branchings are respiratory bronchioles (diameter of about 1 mm) with some gas exchange. The final 4 generations of branches form the alveolar ducts, which terminate in the alveoli. Here, most of the gas exchange takes place: the alveolar ducts and sacs are fully alveolated, while the respiratory bronchioles have variously spaced alveoli. Altogether, between 200 and 600 million alveoli provide a grown person with about 70 to 90 m^2 of exchange surface.

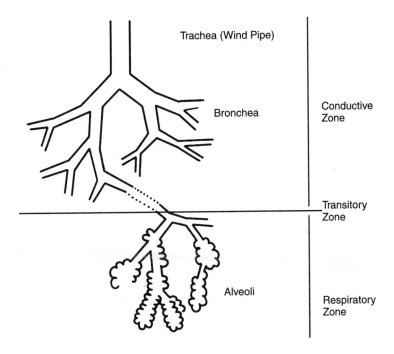

Figure 6-2. Scheme of the respiratory tree.

   The air exchange is brought about by the pumping action of the thorax. The diaphragm separating the chest cavity from the abdomen descends about 10 cm when the abdominal muscles relax (which brings about a protrusion of the abdomen). Furthermore, muscles connecting the ribs contract and, by their anatomical and mechanical arrangements, raise the ribs. Hence, the dimensions of the rib cage and of its included thoracic cavity increase both toward the outside and in the direction of the abdomen: air is sucked into the lungs. When the inspiratory muscles relax, elastic recoil in lung tissue, thoracic wall, and abdomen restores their resting positions without involvement of expiratory muscles: air is expelled from the lungs. However, when ventilation needs are several times higher than the resting value, such as with heavy work, the recoil forces are augmented by activities of the expiratory muscles. These internal intercostal muscle fibers have a direction opposite to those of the external intercostal muscles and reduce the thoracic cavity. Furthermore, contraction of the muscles in the abdominal wall can also assist expiration.
   Thus, inspiratory and expiratory muscles are activated reciprocally, and both overcome the resistance provided by the elastic properties of the chest wall and of pulmonary tissue, particularly of the airways. Of the total resistance, approximately four-fifths is airway resistance and the remainder tissue resistance. The high airway resistance results from the turbulence of the airflow in the trachea and

the main bronchi, particularly at the high flow velocities required by heavy exercise. However, in the finest air tubes, of which there are very many—and, taken together, they have a large volume—airflow is slow and laminar. Altogether, the energy required for breathing is relatively small, amounting to only about 2% of the total oxygen uptake of the body at rest and increasing to not more than 10% at heavy exercise.

FUNCTIONS

Of course, the primary task of the respiratory system is to move air through the lungs, which provide exchange of gases, heat, and water. The respiratory system also conditions the inspired air, adjusting the temperature of the inward flowing air to body temperature, moistening or drying the air, and cleansing it of particles. All this takes place at mucus-covered surfaces in the nose, mouth, and throat. The temperature regulation is so efficient that the inspired air is at about body core temperature, 37°C, when it reaches the end of the pharynx, whether inspiration is through the mouth or nose.

In a normal climate (see Chap. 9), about 10% of the total heat loss of the body, whether at rest or work, occurs in the respiratory tract. This percentage increases to about 25% at outside temperatures of about $-30°C$. In a cold environment, heating and humidifying the inspired air cools the mucosa; during expiration, some of the heat and water is recovered by condensation from the air to be exhaled—hence the runny nose in cold air. Thus, the respiratory tract not only conditions the inhaled air but also recovers some of the spent energy when the air is exhaled.

The respiratory tract also cleans the air of particles of foreign matter. Particles larger than about 10 μm are trapped in the moist membranes and small hairs of the nose. Smaller particles settle in the walls of the trachea, the bronchi, and the bronchioles. Thus, the lungs are kept practically sterile, and an occasional sneeze or cough (with air movements at approximately the speed of sound in the deeper parts of the respiratory system) helps to expel foreign particles.

Respiratory Volumes

The volume of air exchanged in the lungs depends largely on muscular activities, that is, on the requirements associated with the kind of work performed. When the respiratory muscles are relaxed, there is still air left in the lungs. A forced maximal expiration reduces this amount of air in the lungs to the so-called *residual volume* (or residual capacity); see Figure 6-3. The following maximal inspiration adds the volume called *vital capacity*. Both volumes together are the total lung capacity. During rest or submaximal work, only the so-called *tidal volume* is moved, leaving both an inspiratory and an expiratory reserve volume within the vital capacity. (Dead volume, or dead space, is the volume—about 0.2 L—of the conductive zone of the human airways where the air does not come in contact with alveoli and, therefore, does not contribute to the gas exchange.)

Vital capacity and other respiratory volumes are usually measured with the help of a spirometer. The results depend on the age, training, sex, body size, and

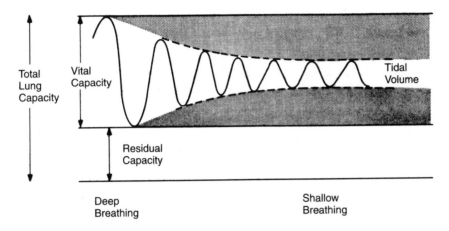

Figure 6-3. Respiratory volumes.

body position of the subject. Total lung volume for highly trained, tall young males is between 7 and 8 L, with a vital capacity of up to 6 L. Women have lung volumes that are about 10% smaller than those of their male peers. Untrained persons have volumes of about 60% to 80% of their athletic peers.

Pulmonary ventilation is the movement of gas in and out of the lungs. The pulmonary ventilation is calculated by multiplying the frequency of breathing by the expired tidal volume. This is called the (respiratory expired) *minute volume*. At rest, we breathe 10 to 20 times per minute. In light exercise, it is primarily the tidal volume that is increased. With heavier work, the respiratory frequency also quickly increases up to about 45 breaths/min (Children breathe much faster at maximal effort—about 55 breaths/min at 12 years of age and about 70 breaths/min at 5 years.) This indicates that breathing frequency, which can be measured easily, is not a very reliable indicator of the heaviness of work performed.

Table 6-1 presents an overview of gas partial pressures at rest, under certain simplifying assumptions, such as completely dry inspired air. The efficiency of gas exchange in the lungs is indicated by the decreased partial oxygen pressure and increased partial pressure of CO_2 in the expired air compared to the inhaled air. The gas pressures in the venous blood depend very greatly on the metabolic activities of the tissues. For example, relatively more oxygen is used in the muscle and more carbon dioxide produced than in other tissues, such as kidneys and skin. This becomes apparent as we compare the gas tensions in venous blood that is either mixed after having passed through various tissues or observed after passage through a muscle. Of course, the gas tensions shown in this table for resting individuals change greatly during strenuous work. For example, the carbon dioxide tension in mixed venous blood at maximal exercise may be in the neighborhood of 70 mm Hg compared to 47 mm Hg at rest; accordingly, the partial pressure of oxygen may be reduced from 40 to 17 mm Hg.

Table 6-1. Gas partial pressures at rest, at 760 mm Hg, with dry inspired air. (Based on Astrand and Rodahl 1977, Rodgers 1986).

	Air			*Blood*		
	Inspired	**Alveolar**	**Expired**	**Arterial**	**Muscle Capillary**	**Mixed Venous**
O_2	150	100	116	95	30	40
CO_2	0.3	40	32	40	50	47
N_2	610	573	565	578	573	576
H_2O	0	47	47	47	47	47

Compared to rest, the respiratory system is able to increase its moved volumes and absorbed oxygen by large multiples. The minute volume can be increased from about 5 L/min to 100 or more L/min, an increase in air volume by a factor of 20 or more. Though not exactly linearly related to it, the oxygen consumption shows a similar increase.

Measurement Opportunities

The changes in pulmonary functions can be used to make assessments of a person's strain while performing physical work: breathing rate, tidal volume, and minute volume are of particular interest. Of these, breathing rate is easily recorded with a plethysmograph, but the rate does not vary much from rest to light work; use of an air flowmeter is feasible as well (Caretti et al., 1994). Yet these measurements provide less information than can be gathered from recordings of circulatory events (especially heart rate) and metabolic functions (especially oxygen consumption). Therefore, measurements of breathing rates and volumes by themselves are not generally performed.

SUMMARY

Air entering the respiratory tree is conditioned so that gas exchange at the alveoli is facilitated: oxygen is absorbed into the blood, and carbon dioxide, heat, and water are dispelled into the air to be exhaled.

Only a part (the tidal volume) of the available volume (vital capacity) of the airways is normally utilized for respiratory exchange.

The respiratory system can easily increase its moved air volume by a factor of 20 over resting conditions, which is accompanied by a similar increase in oxygen intake needed to perform physically demanding work.

At rest, the breathing frequency is less than 20/min. With light work, the frequency remains near resting level, but the tidal volume is larger. In heavy work, breathing rate and tidal volume are augmented.

REFERENCES

Astrand, P.O., and Rodahl, K. (1977). *Textbook of Work Physiology,* 2nd. ed. New York, NY: McGraw-Hill.

Caretti, D.M., Pullen, P.V., Premo, L.A., and Kuhlmann, W.D. (1994). Reliability of Respiratory Inductive Plethysmography for Measuring Tidal Volume During Exercise. *American Industrial Hygiene Association Journal* 55(10):918–923.

Rodgers, S.R. (1986). Personal communication, 24 March 1986.

FURTHER READING

Astrand, P.O., and Rodahl, K. (1986). *Textbook of Work Physiology,* 3rd ed. New York, NY: McGraw-Hill.

Guyton, A.C. (1979). *Physiology of the Human Body,* 5th ed. Philadelphia, PA: Saunders.

Schmidt, R.F. and Thews, G. (eds.) (1989). *Human Physiology* (2nd ed.). Berlin, Germany; New York, NY: Springer.

Stegemann, J. (1981). *Exercise Physiology.* Chicago, IL: Yearbook Medical Publications, Thieme.

Weller, H., and Wiley, R.L. (1979). *Basic Human Physiology.* New York, NY: Van Nostrand Reinhold.

Chapter 7

The Circulatory System

OVERVIEW

Two systems transport materials between body cells and tissues: the blood and the lymphatic system. They take nutritional materials from the digestive tract to the cells for catabolism, synthesis, and deposit. The blood provides oxygen by carrying it from the lungs to the consuming cells, and it carries metabolites: carbon dioxide is brought to the lungs to be expelled, lactic acid is transferred to the liver and kidneys for processing, and heated body fluids, mainly water, are taken to the skin and lungs for heat dissipation. Furthermore, blood is part of the body control system, carrying hormones from the endocrine glands to the cells.

THE MODEL

The circulatory system carries oxygen from the lungs to the cells, where nutritional materials, also brought by circulation from the digestive tract, are metabolized. Metabolic by-products (carbon dioxide, heat, and water) are dissipated by circulation. The circulatory and respiratory systems are closely interrelated, as shown earlier in Figure 6-1.

INTRODUCTION

Blood circulation is essential for providing nutrients and oxygen to body organs, such as muscles; and to remove metabolic by-products. The flow is accomplished by the pumping heart, whose beat rate is often taken as the indicator of the work load.

BODY FLUIDS

Water is the largest weight component of the body: about 60% of body weight in men, about 50% in women. In slim individuals, the percentage of total water is higher than in obese persons since adipose tissue contains very little water. The relation between water and lean (fat-free) body mass is rather constant in normal adults, at about 72%. Of the total body water (say, 40 L in an adult), about 70% (25 L) is contained within body cells (intracellular fluid); separated by the cell membrane, the other 30% (about 15 L) surrounds the cells (extracellular fluid). Extracellular fluid has an ionic composition similar to seawater. The concentrations of various electrolytes are different in the extra- and intracellular fluids. The cell membrane, which separates those two fluids, acts as a barrier for positively charged ions (cations).

Of the extracellular fluid, some is contained within blood vessels (intravascular), while the rest is between the blood vessels and the cell membrane (interstitial fluid); the cells "bathe" in interstitial fluid. Substance exchange between the intravascular and interstitial fluids takes place primarily through the walls of the capillaries; plasma and blood corpuscles, however, cannot pass through these walls.

Blood

Approximately 10% of the total fluid volume consists of blood. The volume of blood is variable, however, depending on age, gender, and training. Volumes of 4 to 4.5 L of blood for women and 5 to 6 L for men are normal. Of these volumes, about two-thirds is usually located in the venous system, with the rest in the arterial vessels. The specific heat of blood is 3.85 J (0.92 cal) per gram.

Of the total blood volume, about 55% is plasma, which consists to $^1/_{10}$ of solids and to $^9/_{10}$ of water. The remaining 45% of the blood volume consists of formed elements. These are predominantly red cells (erythrocytes), white cells (leukocytes), and platelets (thrombocytes). The percentage of red cell volume in the total blood volume is called hematocrit.

Loss of blood is not critical for a healthy person if it is less than 10% of the total blood volume. If the loss approaches 20%, blood pressure is reduced, and changes in pulse and breathing set in. Blood loss of 40% or more means that death is imminent if blood and fluids are not infused.

Blood Groups

According to the content of certain antigens and antibodies, blood is classified into four groups: A, B, AB, O. The importance of these classifications lies primarily in their incompatibility reactions in blood transfusions. However, there are other subdivisions, in particular, the one according to the rhesus (Rh) factor. This is important as an obstetrical problem, which may occur when a pregnant rhesus-negative woman is carrying the child of a rhesus-positive father.

Functions

The plasma carries dissolved materials to the cells, particularly oxygen and nutritive materials (monosaccharides, neutral fats, amino acids—see Chap. 8), as well as hormones, enzymes, salts, and vitamins. Plasma removes waste products, particularly dissolved carbon dioxide and heat.

The red blood cells perform the oxygen transport. Oxygen attaches to hemoglobin, an iron-containing protein molecule of the red blood cell. Each molecule of hemoglobin contains four atoms of iron, which combine loosely and reversibly with four molecules of oxygen. Carbon dioxide molecules can be bound to amino acids of this protein; since these are different binding sites, hemoglobin molecules can react simultaneously with oxygen and carbon dioxide. (Hemoglobin has a strong affinity for carbon monoxide, which takes up spaces otherwise filled by oxygen; this explains the high toxicity of CO.) Each gram of hemoglobin can combine, at best, with 1.34 mL of oxygen. With 150 g of hemoglobin per L of blood, a fully O_2-saturated liter of blood can carry 0.20 L of O_2; also, about 0.003 L of O_2 is dissolved in the plasma. The approximately 5 L of blood in the body circulate through it in about 1 min when the body is at rest.

THE LYMPHATIC SYSTEM

The lymphatic system is separate from the blood vessels. Lymphatic capillaries are found in most tissue spaces. They combine into larger and larger vessels that finally lead to the neck, where they empty into the blood circulation at the juncture of the left internal jugular and left subclavian veins. Thus, the lymphatic system is an accessory to the blood flow system.

The lymphatic capillaries are so small and thin-walled and, thus, so permeable that even very large particles and protein molecules can pass directly into them. Hence, the fluid in the lymphatic system is really an overflow from tissue spaces; lymph is very much the same as interstitial fluid. The flow of lymph within its tubing system is dependent on the interstitial fluid pressure: the higher the pressure, the greater the lymph flow. The second factor affecting lymph flow is the so-called lymphatic pump. After a lymph vessel is stretched by excess lymph, it automatically contracts. This contraction pushes the lymph past lymphatic valves, which allow flow only in the direction of circulation and not backward toward the tissue. The contractions occur periodically, about once every 6 to 10 sec. Finally, lymph can also be pumped by the motion of tissues surrounding the lymph vessels, such as the contraction of skeletal muscle surrounding a vessel.

These lymph-pumping mechanisms, in combination, generate a partial vacuum in the tissues so that excess fluid can be collected and returned through the lymphatic system to the circulation. Lymph flow is highly variable, on the average 1 to 2 mL/min, a small but sufficient amount to drain excess fluid and especially excess protein that otherwise would accumulate in the tissue spaces.

Swelling of the lower legs and feet in the course of long sitting with little motion (for example, during long airplane flights) is a common example of the collection of fluids, particularly in the lymphatic system.

THE CIRCULATORY SYSTEM OF THE BLOOD

Working muscles and other organs requiring blood supply, removal of metabolites, and control through the hormonal system are located throughout the body. Hence, very different and huge demands are placed on the transport system, the circulatory system of the blood.

Architecture of the Circulatory System

It is convenient, as well as anatomically and functionally correct, to model human circulation as a closed loop with two fluid pumps in series: the right and left heart halves. After passing a pump, the blood flows within ever-branching and ever-narrower arteries until it slowly moves through a delicate network of fine vessels, the "vascular bed" of an organ. Here, oxygen and other carried goods are exchanged with the surrounding tissues. This accomplished, the blood drains into small vessels that combine to form the veins, ending in the atrium of the next half of the heart.

The *systemic* subsystem is powered by the left half of the heart. It carries oxygen-rich blood to the capillary beds, traversing metabolizing organs such as the skeletal muscles. There it releases oxygen and energy carriers, such as glucose and glycogen. Then it picks up metabolic by-products, especially carbon dioxide, water, and heat. It drains into the venules of the *pulmonary* subsystem, which is powered by the right heart. The blood then reaches the capillary beds of the lungs, where it releases the metabolic by-products and picks up oxygen.

The total vascular system consists of a large number of parallel, serial, and often interconnected circulatory sections that supply individual organs.

The Heart as Pump

The heart is in essence a hollow muscle that produces, through contraction and with the aid of valves, the desired blood flow. Each half of the heart has an antechamber (atrium) and chamber (ventricle), the pump proper. The atria receive blood from veins, which then flows through open valves into the ventricle. The valves close while the ventricle is compressed by its surrounding musculature.

The mechanisms for excitation and contraction of the heart muscle are quite similar to those of skeletal muscle; however, specialized cardiac cells in the atrium (the sinoatrial nodes) serve as "pacemakers," which do not need external nervous impulses to function. They determine the frequency of contractions by propagating stimuli to other cells in the ventricle, especially Purkinje's fibers, which make the ventricular muscles contract.

The heart's own intrinsic control system operates, without external influences, at rest with (individually different) 50 to 70 beats per minute (bpm). Changes in heart action stem from the central nervous system, which influences

the heart through the sympathetic and the parasympathetic subsets of the autonomic system (see Chap. 4). Stimulation toward increase of heart action comes
through the sympathetic system, mostly by increasing the heart rate, the strength
of cardiac contraction, and the blood flow through the coronary blood vessels
supplying the heart muscle. The parasympathetic system causes a decrease
in heart activities, particularly a reduced heart rate, weakened contraction of the
atrial muscle, slowed conduction of impulses (lengthening the delay between
atrial and ventricular contraction), and decreased blood flow through the coronary blood vessels. The parasympathetic system is dominant during rest periods.
Thus, the sympathetic and parasympathetic nervous systems are another example of the coordinated action of two opposing control systems, so often found in
the body.

The myocardial action potentials can be recorded in their algebraic sum by
an electrocardiogram (ECG or EKG). The different waves observed in the ECG
have been given alphabetic identifiers: the P wave is associated with the electrical
stimulation of the atrium, while the Q, R, S, and T waves are associated with
ventricular events. The ECG is mostly employed for clinical diagnoses; however, with appropriate apparatus, it can be used for counting and recording the
heart rate. Figure 7-1 shows the electrical, pressure, and sound events during a
contraction-relaxation cycle of the heart.

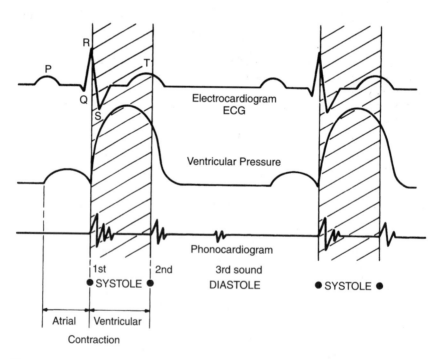

Figure 7-1.   Scheme of the electrocardiogram, the pressure fluctuation, and the
phonogram of the heart with its three sounds (adapted from Guyton, 1979).

The ventricle is filled through the valve-controlled opening from the atrium. During systole, when the heart muscle contracts and when the internal pressure is equal to the pressure in the aorta, the aortic valve opens and the blood is ejected from the heart into the systemic system if the left side of the heart is involved and into the pulmonary system if the right heart is considered. Continuing contraction of the heart further increases the pressure since less blood can escape from the aorta than the heart presses into it. Part of the excess volume is kept in the aorta and its large branches, which together act as a *windkessel,* an elastic pressure vessel. Then, the aortic valve closes with the beginning of the relaxation (*diastole*) of the heart, and the elastic properties of the aortic walls propel the stored blood into the arterial tree, where the elastic blood vessels smooth out the waves of blood volume.

At rest, about half the volume in the ventricle is ejected (stroke volume), while the other half remains in the heart (residual volume). Under exercise load, the heart ejects a larger portion of the contained volume. When much blood is required but cannot be supplied, such as during very strenuous physical work with small muscle groups or during maintained isometric contractions, the heart rate can become very high.

At a heart rate of 75 bpm, the diastole takes less than 0.5 sec and the systole just over 0.3 sec. At a heart rate of 150 bpm, the periods are close to 0.2 sec each. Hence, an increase in heart rate occurs mainly by shortening the duration of the diastole.

The events in the right heart are similar to those in the left, but the pressures in the ventricular and pulmonary arteries are only about one-fifth of those during systole in the left heart.

Cardiac Output

The cardiac output can be affected by two factors: the frequency of contraction (heart rate) and the pressure generated by each contraction in the blood. Both determine the so-called (cardiac) minute volume. (The available blood volume does not vary.) The cardiac output of an adult at rest is around 5 L/min. When performing strenuous exercise, this level might be raised fivefold to about 25 L/min, while a well-trained athlete may reach up to 35 L/min. The ability of the heart to adjust its minute output volume to the requirements of the activity depends on two factors: the effectiveness of the heart as a pump and the ease with which blood can flow through the circulatory system and return to the heart. A healthy heart can pump much more blood through the body than is usually needed. Hence, an output limitation is more likely to lie in the transporting capability of the vascular portions of the circulatory system than in the heart itself. Of the total vascular system, the arterial section (before the metabolizing organ) has relatively strong elastic walls, which act as a pressure vessel transmitting pressure waves far into the body, though with much loss of pressure along the way. At the arterioles of the consumer organ, the blood pressure is reduced to approximately one-third its value at the heart's aorta. Figure 7-2 shows schematically the smoothing of pressure waves and reduction of pressure along the blood pathway. As blood seeps through the consuming organ (such as a muscle) via capillaries, pressure

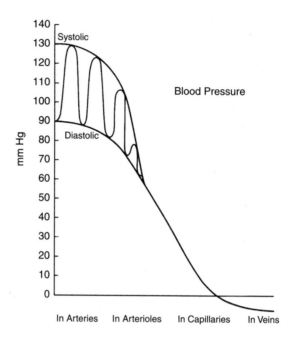

Figure 7-2. Scheme of the smoothing and reduction of blood pressure along the circulatory pathways, showing the blood pressure differential between arterial and venous sections.

drops. The pressure differential (positive on the arterial side, negative on the venous side) helps to maintain blood transport through the capillary bed.

The Capillary Bed

The two sides of the vascular beds (in the lungs or other organs) have similar architectural patterns. In the direction of the blood flow, the blood vessels branch so prolifically that their total volume increases even as the cross-section area decreases. (Noordergraaf (1978) estimates the number of capillaries to exceed 10^9 in the human.) This slows the blood flow in the bed and allows diffusion through the large area of thin cell walls. On the exit side of the capillary bed, the many small vessels combine to form fewer but larger vessels. Oxygen enters the bloodstream in the pulmonary capillaries with an operational area of 90 m^2 and is delivered through the systemic capillaries with a surface of 200 m^2, as estimated by Noordergraaf.

A capillary bed is drawn schematically in Figure 7-3. Blood enters through the arteriole, which is circumvoluted by smooth muscles that contract or relax in response to stimuli from the sympathetic nervous system and to local accumulation of metabolites. The following metarteriole has fewer enclosing muscle

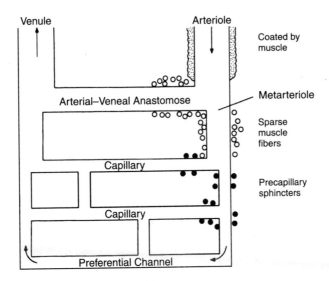

Figure 7-3. The capillary bed, showing the branching connections between arterial and venous parts of the circulatory system and the smooth sphincter muscles that bring about vascular constriction.

fibers, and the entrance to the capillaries may be closed by other ringlike muscles, the precapillary sphincters. The metarteriolic and the precapillary sphincter muscles are predominantly controlled by local tissue conditions. Contraction or relaxation of the flow-controlling muscles changes the flow resistance and, hence, the blood pressure. If lack of oxygen or accumulation of metabolites requires high blood flow, the muscles allow the pathways to remain open; blood may even use a shortcut established by an arteriole-venule shunt (anastomosis). Large cross-sectional openings in the system reduce blood flow velocity and blood pressure, allowing nutrients and oxygen to enter the extracellular space of the tissue and permitting the blood to accept metabolic by-products from the tissue. (However, if contraction of striated muscle compresses the fine blood vessels, flow may be hindered or shut off. This is of particular consequence in sustained, strong isometric contraction; see Chap. 3.) Constriction of the capillary bed reduces local blood flow so that other organs in greater need of blood may be better supplied.

The venous portion of the systemic system has a large cross section and provides low flow resistance; only about one-tenth of the total pressure loss occurs here. (This low pressure system is often called the capacitance system in contrast to the arterial resistance system.) Valves are built into the venous system, allowing blood flow only toward the right ventricle. The pulmonary circulation has relatively little vessel constriction or shunting.

Hemodynamics

As in the physics of classical fluid dynamics, the important factors in the dynamics of the blood flow (hemodynamics) are the capacity of the pump (the heart) to do the work, the physical properties of the fluid (blood) to be pumped, particularly its viscosity ("internal friction"), and the properties of the pipes (blood vessels) with respect to the required flow rates (volume per unit time) and flow velocities.

The pressure gradient (the difference in internal pressure between the start and the end of the flow pathway considered) is the main determiner of flow rate and flow velocity but depends on the resistance to flow. The equation

$$\Delta p = QR \qquad\qquad (7\text{-}1)$$

describes the relation between the pressure gradient $\Delta p = (p_{start} - p_{end})$, the flow rate Q and the peripheral resistance R.

Since the pressure in a fluid is (theoretically), at any given point, equal in all directions, it can be measured as the pressure against the lateral walls of the blood vessel. This lateral pressure is usually measured as "blood pressure," with its reading modified by the elastic properties of the containing blood vessel and, perhaps, by other intervening tissues.

The flow resistance of the blood vessels is quite different from the formula used for rigid pipes but depends in essence on the diameter and length of the vessel considered and on the viscosity of the blood. Blood viscosity is about three to four times greater than that of water, determined primarily by the number, dimensions, and shapes of the blood cells and dependent on the protein content of the plasma; the more hematocrit, the higher the resistance to flow.

The stream of blood may not be streamlined (laminar) but turbulent. Such turbulence can result from the fact that the outer layers of the bloodstream are in physical contact with the inner walls of the blood vessels and have, therefore, a much lower velocity than the more central sections of the blood flow. Turbulence increases the energy loss within the energy medium and, therefore, at a given pressure gradient, the average flow is slower.

The static pressure in a column of fluid depends on the height of that column (Pascal's law). However, the hydrostatic pressure in, for example, the feet of a standing person is not as large as expected from physics since the valves in the veins of the extremities modify the value: in a standing person, the arterial pressure in the feet may be only about 100 mm Hg higher than in the head. Nevertheless, blood, water, and other body fluids in the lower extremities are pooled there, leading to a well-known increase in volume of the lower extremities, particularly when one stands (or sits) still for a long enough time.

Blood Vessels

Of all the blood vessels, the capillaries and, to some extent, also the postcapillary venules provide semipermeable membranes to the surrounding tissue so that nutrients, gases, and so forth, may be exchanged. All other blood vessels—the arterioles, arteries, and veins—serve only as transport channels. The walls of these vessels consist (in different compositions) of elastic fibers, collagen fibers, and smooth muscles. The walls are thick in the big arteries and thin in the big veins. Blood flow at each point in these blood vessels is determined primarily by the pressure head of the blood wave and the diameter of the vessel at this point. Different vessels have varying capacities to influence the blood flow, blood distribution, and vascular resistance.

As already discussed, the arteries serve as pressure tanks during the ejection of the blood from the heart. Their elastic tissues stretch under the systolic impact, store this energy, and release it during diastole, thus converting an intermittent flow to a much more continuous stream. Still, the ejection of the blood from the left ventricle causes a pressure wave to travel along the blood vessels at speeds of 10 to 20 times the velocity of the blood in the aorta, which is about 0.5 m/s at rest. The frequency at which these pressure waves occur is counted as pulse rate or heart rate.

The flow resistance in the large arteries and veins is very small since the diameter of these vessels is large and the flow velocity high. In contrast, the peripheral resistance in the arterioles, metarterioles, and capillaries is high, which causes a significant reduction in blood pressure, even though the velocity of the (turbulent) blood flow is still high. The diameter of the vessels, below 0.1 mm, can be further reduced by smooth muscle fibers that are wrapped around the vessel. Contraction of these muscles constricts the vessels, which recoil to their normal size when the muscle relaxes. Thus, the contraction of these smooth muscles around the blood vessels (and possibly the pressure generated by contraction of surrounding striated muscle) can change the flow characteristics and achieve reduction or complete shutoff of circulation to organs where blood supply is not so urgent, allowing better supply to those organs in need of it. Any increase in channel capacity (vasodilation) must be compensated for by vessel constriction in other areas or by an increase in cardiac output, or both, since, otherwise, the arterial blood pressure would fall. In the capillaries, there is no pressure variation associated with the heart's systolic and diastolic phases.

The collecting venules have a coating of connective tissue and smooth muscle rings, only intermittently spaced near the arterial bed but developing into a complete layer in their distal parts, the larger veins. Valves in the veins of extremities prevent backflow of the blood. Compression of skeletal muscle reduces the volume of veins, which is restored during relaxation. Given the high flow resistance on the capillary side and the design of the valves in the veins, the blood is squeezed toward the right ventricle. Thus, the alternately contracting and relaxing muscle (doing dynamic work) acts as a pump that moves venous blood toward the right heart. The venous system usually carries about two-thirds of the total blood volume; of this, the pulmonary veins contain about 15% of the total blood volume.

REGULATION OF CIRCULATION

If organs such as muscles need increased blood flow, flow regulation takes place primarily at two sites: the heart and the organ to be supplied.

The blood pressure in the aorta depends on cardiac output, peripheral resistance, elasticity of the main arteries, viscosity of the blood, and blood volume.

The local flow is determined mainly by the pressure head and the diameter of the vessel through which it passes. The smooth muscles encompassing the arterioles and veins continuously receive sympathetic nerve impulses that keep the clearance (lumen) of the vessels more or less constricted. This local vasomotor tone is controlled by the vasoconstricting fibers driven from the medulla; the tone keeps the systemic arterial blood pressure on a level suitable to the actual requirements of all vital organs. (Changes in heart function and in circulation are initiated at brain levels above the medullary centers, probably at the cerebral cortex.) If local metabolite concentrations increase over an acceptable level, this local condition directly causes metarteriolic and sphincter muscles to relax, allowing more blood flow. In particular, if muscles require more blood flow in heavy work, signals from the motor cortex can activate vasodilation of precapillary vessels in the muscles and simultaneously trigger a vasoconstriction of the vessels supplying the abdominal organs. This leads to a remarkable and very quick redistribution of the blood supply to favor skeletal muscles over the digestive system; this is called the muscles-over-digestion principle.

Even with heavy exercise, however, the systemic blood flow is so controlled that the arterial blood pressure is sufficient for an adequate blood supply to the brain, heart, and other vital organs. To bring this about, neural vasoconstrictive commands can override local dilatory control. For example, the temperature-regulating center in the hypothalamus can affect vasodilation in the skin if this is needed to maintain a suitable body temperature, even if it means a reduction in blood flow to the working muscles—the skin-over-muscles principle (see Chap. 9).

Thus, on the arterial side, circulation at the organ/consumer level is regulated both by local control and by impulses from the central nervous system, the latter having overriding power. Vasodilation (opening blood vessels beyond their lumen at regular "vasomotor tone") in the organs needing increased blood flow, and vasoconstriction where blood is not so necessary, regulate local blood supply. At the same time, the heart increases its output by higher heartbeat frequency; the blood pressure also increases. At the venous side of circulation, constriction of veins, combined with the pumping action of dynamically working muscles and the forced respiratory movements, facilitates return of blood to the heart. This makes increased cardiac output possible because the heart cannot pump more blood than it receives.

Heart rate generally follows oxygen consumption and, hence, energy production of the dynamically working muscle in a linear fashion from moderate to rather heavy work (see "Indirect Calorimetry" in Chap. 8). However, the heart rate at a given oxygen intake is higher when the work is performed with the arms than with the legs. This reflects the use of different muscles and muscle masses with different lever arms to perform the work. Smaller muscles doing the same external work as larger muscles are more strained and require more oxygen.

Also, static (isometric) muscle contraction increases the heart rate, apparently because the body tries to bring blood to the tensed muscles. It is difficult, however, to compare this effect in terms of efficiency (e.g., in beats per effort) with the increase in heart rate during dynamic efforts because, in the isometric case, there is no "work" done (in the physical sense, there is force but no displacement) while, in the dynamic case, work is done.

Of course, work in a hot environment causes a higher heart rate than at a moderate temperature, as explained in Chapter 9. Finally, emotions, nervousness, apprehension, and fear can affect the heart rate at rest and during light work.

Measurement Opportunities

Given current technology, it is impractical to measure the changes in blood supply at the working muscle although this would be of special interest because the local supply largely determines whether the muscle can perform its job. Current instrumentation is capable of easily measuring functions at and near the heart; heart (pulse) rate can be recorded by electrical and volume (pressure) changes without using invasive techniques; and the sound can be heard easily even without a stethoscope.

SUMMARY

The blood flow per minute within the circulatory system depends on the actual operating characteristics of both the heart (as pump) and the blood vessels that contain and guide the flow of the blood. The volume of blood pumped per minute is in the neighborhood of 5 L at rest and may increase fivefold or even sevenfold at strenuous work. This increase is brought about mostly by changes in heart rate and blood pressure per heartbeat.

Flow through the arterial and following venous subsystems is determined by the organs selected by the body for their need of blood and by the local muscle contraction conditions. Table 7-1 indicates changes in blood supply for various body parts at rest and work.

Table 7-1. Blood supply to consuming organs during rest and work (adapted from Astrand and Rodahl 1977).

Consumer	At Rest*	At Heavy Work*
Muscle	15	75
Heart	5	5
Digestive tract	20	3
Liver	15	3
Kidneys	20	3
Skin	5	5
Bone	5	1
Fatty tissues	10	1
Brain	5	5

*estimated percent of the cardiac output which is about 5 L min^{-1} at rest, and 25 L min^{-1} at heavy work.

Oxygen intake ("uptake") is the volume of oxygen (at defined atmospheric conditions) absorbed per minute from the inspired air. At rest, this volume is about 0.2 L/min, which can increase at strenuous exercise up to thirtyfold.

Heart rate (HR) is the number of ventricular contractions per minute. Its range is usually 60 to 70 bpm at rest and increases threefold at strenuous exercise. Pulse rate is the frequency of pressure waves in the arteries and is the same as heart rate in normal, healthy individuals.

Stroke volume (SV) is the volume of blood ejected from the left heart into the main artery during each ventricular contraction. It is usually 40 to 60 mL at rest and may increase threefold with hard work.

Cardiac output, also called cardiac minute volume, is the volume of blood injected into the main artery per minute: stroke volume multiplied with heart rate equals cardiac output. It may increase five- to sevenfold over resting values.

Blood pressure (BP) is the internal pressure in the arteries near the heart, at rest about 70 mm Hg during diastole and 120 during systole. These values may double with heavy exercise.

REFERENCES

Astrand, P.O., and Rodahl, K. (1977 and 1986). *Textbook of Work Physiology,* 2nd and 3rd eds. New York, NY: McGraw-Hill.

Guyton, A.C. (1979). *Physiology of the Human Body,* 5th ed. Philadelphia, PA: Saunders.

Noordergraaf, A. (1978). *Circulatory System Dynamics.* New York, NY: Academic Press.

Schmidt, R.F. and Thews, G. (eds.) (1989). *Human Physiology* (2nd ed.). Berlin, Germany; New York, NY: Springer.

Chapter 8

The Metabolic System

OVERVIEW

Over time, the human body maintains a balance (homeostasis) between energy input and output. The input is determined by the nutrients, from which chemically stored energy is liberated during the metabolic processes within the body. The output is mostly heat and work, with work measured in terms of physically useful energy, that is, energy transmitted to outside objects. The amount of such external work performed strains individuals differently, depending on their physique and training.

THE MODEL: THE "HUMAN ENERGY MACHINE"

Astrand and Rodahl (1977) used an analogy between the human body and an automobile. In the cylinder of the engine, an explosive combustion of a fuel-air mixture transforms chemically stored energy into physical kinetic energy and heat. The energy moves the pistons of the engine, and gears transfer their motion to the wheels of the car. The engine needs to be cooled to prevent overheating. Waste products are expelled. This whole process can work only in the presence of oxygen and when there is fuel in the tank. In the human machine, muscle fibers are both cylinders and pistons: bones and joints are the gears. Heat and metabolic by-products are generated while the muscles work. Nutrients (mostly carbohydrates and fats) are the fuels that must be oxidized to yield energy.

INTRODUCTION

This chapter first provides an overview of human metabolism and work and then discusses, in three sections, the process of energy liberation, the assessment of a person's capacity for energy expenditure, and the energy requirements at work. Figure 8-1 shows schematically the main interactions within the body.

Human Metabolism and Work

The term *metabolism* includes all chemical processes in the living body. In a narrower sense, it is often used—as it is here—to describe the (overall) energy-yielding processes.

The balance between energy input I (via nutrients) and outputs can be expressed by an equation[1]:

$$I = M = H + W + S \qquad (8\text{-}1)$$

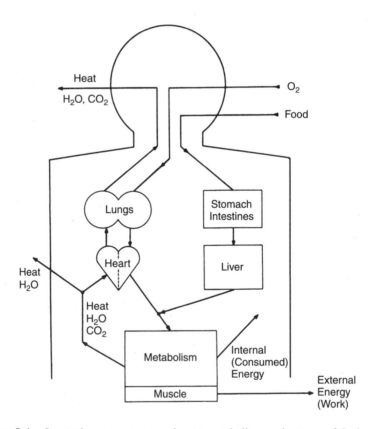

Figure 8-1.   Interactions among energy inputs, metabolism, and outputs of the human body.

---

[1]This terminology is slightly different from that used in the previous two editions of this book.

where M is the metabolic energy generated, which is divided into the heat H that must be dispelled to the outside, the work W performed, and the energy storage S in the body (positive if it increases, negative if it decreases).

The measuring units for energy (work) are joules (J) or calories (cal) with 4.2 J = 1 cal. (Exactly: 1 J = 1 Nm = 0.2389 cal = 10^7 ergs = 0.948 × 10^{-3} Btu = 0.7376 ft lb.) One uses the kilocalorie (i.e., kcal or Cal = 1,000 cal) to measure the energy content of foodstuffs. The units for power are 1 kcal/hr = 1.163 W, with 1 W = 1 J/s.

Assuming, for simplicity, that there is no change in energy storage and that no net heat is gained from, or lost to, the environment (see Chap. 9 for details about actual heat exchange), we can simplify the energy balance equation as

$$I = H + W \qquad (8\text{-}2)$$

Human energy efficiency (work efficiency) is defined as the ratio between work performed and energy input.

$$e \text{ (in } \%) = \frac{100\ W}{I} = \frac{100\ W}{M} \qquad (8\text{-}3)$$

In everyday activities, only about 5% or less of the energy input is converted into "work," that is, energy usefully transmitted to outside objects; highly trained athletes may attain, under favorable circumstances, perhaps 25%. The remainder (that is, most) of the input is converted into heat, usually at the end of a long chain of internal metabolic processes.

Work (in the physical sense) is done by skeletal muscles (see Chap. 3), which move body segments against external resistances. The muscle is able to convert chemical energy into physical work (energy). From resting, it can increase its energy generation up to fiftyfold. Such enormous variation in metabolic rate not only requires quickly adapting supplies of nutrients and oxygen to the muscle but also generates large amounts of waste products (mostly heat, carbon dioxide, and water), which must be removed. Thus, the ability to maintain the internal equilibrium of the body while it is performing physical work depends largely on the circulatory and respiratory functions that serve the involved muscles. Among these functions, the control of body temperature is of special importance. This function interacts with the external environment, particularly with temperature and humidity (as discussed in Chap. 9).

The work (output) capability of the human body is dependent on its internal ability to generate energy over various time periods at varying energy levels. The engineer determines the "work" required and how it is to be done and, thus, largely has control over the external environment. To arrange for a suitable match between capabilities and demands, the engineer needs to adjust the work to be performed (and the work environment) to the body's energetic capabilities. These human capabilities are determined by the individual's capacity for energy output (physique, training, health), the neuromuscular function characteristics (such as muscle strength and coordination of motion), and psychological factors (such as motivation).

SECTION 1. ENERGY LIBERATION IN THE HUMAN BODY

Overview

Energy is introduced into the body through the mouth in the form of food and drink. The main energy carriers are carbohydrates, fats, and proteins. After passing through the stomach, the nutrients are absorbed in the small intestines into blood or lymph. The absorbed nutrients are then assimilated and stored either as glycogen or as (tissue-building) fat, or they are immediately used to provide energy for the body. At the mitochondria of muscles, they supply the energy needed to convert ADP (adenosine diphosphate) into ATP (adenosine triphosphate); this is the energy source for muscular work.

Energetic Reactions

Energy transformation in living organisms involves chemical reactions that either liberate energy, most often as heat (such reactions are called exergonic or exothermic) or require energy input (these are called endergonic or endothermic reactions).

Generally, breakage of molecular bonds is exergonic, while formation of bonds is endergonic. Depending on the molecular combinations, bond breakage yields different amounts of released energies. Often, reactions do not simply go from the most complex to the most broken-down state but achieve the process in steps with intermediate and temporarily incomplete stages.

The Pathways of Digestion

Energy is supplied to the body as food or drink taken through the mouth (ingestion). Here, the food is chewed into small particles and mixed with saliva (mastication). Saliva is 99.5% water; the rest is salt, enzymes, and other chemicals. Saliva contains mucus, which makes the particles of food stick together in a convenient size (bolus) and lubricates food for passage (deglutition) down to the stomach. The enzyme lysozyme destroys bacteria that otherwise would attack the mucous membranes and teeth.

During swallowing, breathing stops for a couple of seconds so that the bolus avoids getting into the windpipe (trachea) and can slide down the gullet (esophagus). The slide takes up to 8 sec for boli of formerly solid food but only about 1 sec for liquids.

Churning movements (2–4/min) inside the stomach help the gastric juice to break up the bolus until it is fully liquefied (chyme). Alcohol is mostly absorbed while in the stomach into the bloodstream. With the help of pepsin (an enzyme) and hydrochloric acid, digestion is initiated in the breakdown of proteins, but there is little effect on the molecules of fats and carbohydrates. Fatty food stays in the stomach for up to 6 hr; protein-rich materials are passed more quickly while carbohydrates leave within 2 hr.

Most of the chemical digestion of foods takes place in the duodenum, about the first 25 cm of the small intestine (which at 3 cm has a smaller diameter than the large intestine but at 7 m length is about five times longer). Circular muscles mix its content, while longitudinal muscles contract in waves and propel the content downstream. The inner surface of the small intestine has many fingerlike projections (villi), which increase the contact surface. Blood and lymph capillaries are embedded in the surfaces; digested foods are absorbed into these capillaries. The pancreas adds digestive enzymes (and hormones). The liver and gallbladder add bile, a salt-rich fluid that helps in emulsification and absorption of fats. During the 3 to 5 hr the foodstuffs stay in the small intestine, about 90% of all nutrients are extracted.

The final extraction of nutrients takes place in the large intestine, which has a diameter of about 5 cm and a length of approximately 1.5 m. Again, circular muscles mix, and longitudinal muscles propel, its content. In the first section, the cecum, some chemical digestion is done; in the follwing colon, water and electrolytes are absorbed. After having been shaped into a soft, solid mass (feces), solid wastes and undigested food components are egested (defecated). Nitrogenous wastes are transported by the blood to the kidneys, where they are excreted into the urine. Figure 8-2 shows the path of digestion in a schematic sketch.

Some digestion takes place in the stomach, but most of it in the small intestines. This involves not only a physical reduction of solid to liquefied food, but more importantly changes it chemically by breaking large complex molecules into smaller ones. These can be transported through membranes of the body cells and absorbed. The large variety of chemical reactions that follow on the digested and absorbed foodstuffs is called *assimilation:* a reassembly into molecules that can be either easily degraded to release their energy content, stored as energy reserves, or used as raw materials for body growth and repair. More details are discussed below.

Altogether, it takes 5 to 12 hr after eating to extract the nutrients and energy from food. More time is needed to utilize them in the body.

The Energy Content of Nutrients

Our food consists of various mixtures of organic compounds (foodstuffs) and water, salts and minerals, vitamins, and so forth and of fibrous organic material (primarily cellulose). This roughage, or bulk, improves mechanical digestion by stretching the walls of the intestines but does not release energy.

The primary foodstuffs are carbohydrates, fats, and proteins. Their average nutritionally usable energy contents per gram are: 4.2 kcal (18 kJ) for carbohydrate, 9.5 kcal (40 kJ) for fat, and 4.5 kcal (19 kJ) for protein. Alcohol carries about 7 kcal (30 kJ)/g.

The energy value of foodstuffs is measured in a bomb calorimeter, in which food material is electrically burned so that its content is completely reduced to carbon dioxide (CO_2), water (H_2O), and nitrogen oxides. The heat that is developed is the measure for the energy content of the food. The energy content of our daily food and drink depends on the mixture of the basic foodstuffs therein (see Table 8-1). It is of some interest to note that protein oxidized in the body yields

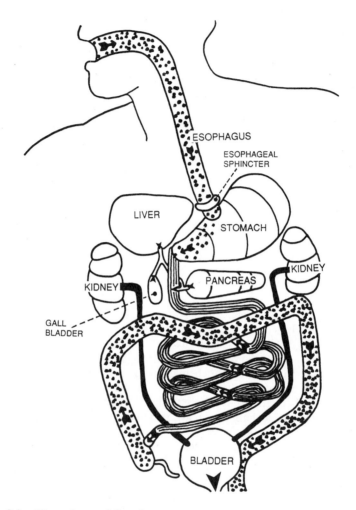

Figure 8-2.   The pathway of digestion.

only 4.5 kcal/g, which is about 75% of the energy freed in the calorimeter. Hence, energy liberation from protein in the body is less efficient than from fats and carbohydrates.

Carbohydrates range from small to rather large molecules, and most are composed of only the three chemical elements carbon, oxygen, and hydrogen. (The ratio of H to O usually is 2:1, just as in water: hence the name carbohydrate, meaning "watered carbon"). Carbohydrates exist as simple sugars (monosaccharides), double sugars (disaccharides), and polysaccharides (which are a large number of monosaccharides joined into chains). The most common natural polysaccharides are plant starch, glycogen, and cellulose. Carbohydrates are digested

Table 8-1.   Estimated energy content of foods and drinks.

Item	Serving size	kcal per serving	kcal/100 g (or 100 mL)	kJ/100 g (or 100 mL)
Meat, Poultry, Fish				
Bacon, fried	(1 slice)	35	585	2440
Chicken, roasted w/o skin	(4 oz)	215	190	795
Hamburger, cooked	(4-oz patty)	300	265	1100
Hot dog, beef	(1)	145	320	1345
Roast beef, lean	(4 oz)	190	170	700
Steak, cooked	(4 oz)	240	210	880
Turkey, light meat	(4 oz)	180	185	655
Halibut, cooked	(4 oz)	195	170	715
Red snapper, raw	(4 oz)	105	95	390
Salmon, cooked	(4 oz)	210	185	760
Shrimp, raw	(4 oz)	105	95	380
Trout (brook), cooked	(4 oz)	225	200	820
Tuna, canned in oil	(4 oz)	240	210	880
canned in water	(4 oz)	125	110	460
Other				
Sugar	(1 cup)	770	385	1610
Honey	(1 tbsp)	60	300	1290
Jam	(1 tbsp)	55	275	1150
Snack Foods				
Brownies	(1)	130	405	1700
Chocolate chip cookies	(4)	185	460	1925
Cupcakes, chocolate icing	(1)	130	360	1510
Hard candy	(1 oz)	110	390	1630
Milk chocolate candy	(1 oz)	160	535	2230
Oatmeal cookies with raisins	(40)	245	470	1965
Potato chips	(10)	115	575	2400
Pretzels	(1 oz)	110	400	1655
Roasted peanuts	(1 cup)	840	585	2445
Beverages				
Cola	(12 fl oz)	145	40	170
Coffee	(12 fl oz)	4	2	8
Hot chocolate	(12 fl oz)	330	85	365
Tea	(12 fl oz)	0	0	0
Milk, whole	(12 fl oz)	200	65	265
Milk, low fat	(12 fl oz)	160	50	205
Milkshakes	(12 fl oz)	350–380	120–130	510–545
Orange or apple juice	(12 fl oz)	160	45	190
Beer: Regular	(12 fl oz)	140–160	40–50	170–190
Light	(12 fl oz)	100–135	30–40	110–160
Wine (red and white)	(3$^1/_2$ fl oz)	80	80	325

Volume units: 16 tbsp = 1 cup = 8 fl oz = 237 mL
Mass units: 1 oz = 28.4 g

Table 8.1. (Continued)

Item	Serving size	kcal per serving	kcal/100 g (or 100 mL)	kJ/100 g (or 100 mL)
Beverages				
Liquor (bourbon, gin, scotch, vodka):				
80 proof	(jigger of	95	225	940
90 proof	1½ fl oz)	110	260	1085
"Fast Foods"				
"Single" hamburger	(100–125 g)	260–340	260–270	1090–1130
"Specialty" hamburger (e.g., Deluxe, Double, Big Mac, Whopper)	(200–280 g)	550–670	240–275	1005–1150
Sandwiches: Chicken	(130–210 g)	320–690	250–330	1045–1380
Fish	(140–205 g)	435–540	265–310	1110–1300
Roast beef	(150 g)	320–350	215–235	900–985
Fried chicken	(40–70 g)	105–200	265–285	1110–1190
French fries, regular	(70 g)	210	300	1255
Cheese pizza, reg. crust	(1/4 of 12-in.-	325	231	970
"Super supreme" pizza	diam pizza)	520–590	390	1650

Volume units: 16 tbsp = 1 cup = 8 fl oz = 237 mL
Mass units: 1 oz = 28.4 g

by breaking the bonds between saccharides so that the compounds are reduced to simple sugars. The monosaccharides produced in the digestion of food carbohydrates are principally glucose (80%); the others are fructose, and galactose; the latter two monosaccharides are quickly converted to glucose, which can be absorbed through the walls of the intestines into the bloodstream.

Fat is called a triglyceride because a fat molecule is formed by joining one glycerol nucleus to three fatty-acid radicals. Unsaturated fat has double bonds between adjacent carbon atoms in one or more of the fatty-acid chains; hence, the compound is not saturated with all the hydrogen atoms it could accommodate. Some of the available bonds are, in fact, not occupied by hydrogen atoms but increase the number of linkages between carbon atoms themselves. The more unsaturated a fat is, the more it is liquid, an oil. Most plant fats are polyunsaturated, while most animal fats are saturated and hence solid. (A diet with a high content of saturated fat is medically suspect.)

Digestion of fat takes place in the small intestine; chemically, it is the breakage of bonds that link the glycerol residue to the three fatty-acid residues. Glycerol and fatty-acid molecules are small enough to cross cell membranes and, hence, can be absorbed. The bloodstream can transport only water-soluble materials, such as glycerol. Many fatty acids are water-repellent: these are absorbed into the lymph vessels (see Chap. 7).

Proteins are chains of amino acids joined together by peptide bonds. Many such different bonds exist and, thus, proteins come in a large variety of types and sizes. During digestion, which occurs partially in the stomach but mostly in the small intestine, the bonded protein is broken into amino acids, which are then absorbed into the bloodstream. The amino acids are transported to the liver, which disperses some of the acids to cells throughout the body to be rebuilt into new proteins. Other amino acids become enzymes, that is, organic catalysts that control chemical reactions between other molecules without being consumed themselves. (Enzymes secreted by the stomach and the pancreas are components of digestive juices and play an important role in chemical digestion.) Still other amino acids become hemoglobin—the oxygen carrier in the blood—or hormones or collagen. Obviously, the body has many important uses for proteins, not only as an energy source.

The Pathways of Absorption and Assimilation

In the stomach, primarily water, salts, and certain drugs (such as alcohol) are absorbed. Most of the digested foodstuffs are assimilated in the small intestine. Here, all digested foodstuffs enter the blood capillaries or, in the case of fatty acids the lymphatic system. The blood capillaries drain eventually into the hepatic portal vein, which carries blood to the liver. This vein also receives inputs from the stomach, pancreas, gallbladder, and spleen. Lymph flows from the intestinal walls through the thoracic duct to the left subclavian vein. Here the lymph enters the bloodstream and becomes part of the blood plasma. The liver receives blood not only from the portal vein but also from the hepatic artery.

Liver cells remove digestion products from the blood and store or metabolize them. Glycogen $(C_6H_{10}O_5)_x$, the storage form of carbohydrate, is generated in the liver from glucose $(C_6H_{12}O_6)$. (This can also be done at muscle cells.) Neutral fat is synthesized in the liver from glucose (one can "get fat" without eating any), fatty acids, and possibly amino acids derived from proteins. This neutral fat then serves as the body's main energy storage. Thus, the liver controls much of the fat utilization of the body.

Assimilation of foodstuffs can occur in two complementary processes. In anabolism (constructive metabolism), small molecules are assembled into larger ones by chemical reactions requiring an input of energy (endergonic metabolism). This needed energy is supplied by catabolism (destructive metabolism), in which organic molecules are broken down, releasing their internal bond energies (exergonic metabolism).

Energy Release

After digestion, absorption, and assimilation, the principal energy carriers are in the body in form of:

- glucose and the closely related glycogen,
- neutral fat,
- protein.

Of these, glucose is the most easily used energy carrier. It can be catabolized in two ways.

Aerobic Metabolism of Glucose

Glucose can be oxidized according to the stoichiometric formula

$$C_6H_{12}O_6 + 6O_2 = 6CO_2 + 6H_2O + \text{energy} \qquad (8\text{-}4)$$

This means that 1 molecule of glucose combines with 6 molecules of oxygen, imported to the process from the outside, and results in 6 molecules each of carbon dioxide and water, while energy (about 690 kcal/mole, or 39 ATP) is released. (Note that equal volumes of water, carbohydrate, and oxygen take part in that conversion. This is important for the measurement of actual energy use in the body, especially with respect to the respiratory quotient, discussed under Indirect Calometry.)

Chemically, oxidation is defined as a loss of electrons from an atom or a molecule (while reduction is defined as a gain of electrons). In such reactions, electrons are carried in the form of hydrogen atoms, and the oxidized compound is dehydrogenated. In human metabolism, organic fuels—glucose, fats and, occasionally, proteins—constitute the major electron donors, while oxygen is the final electron acceptor (oxidant) of the fuel.

Anaerobic Metabolism of Glucose

Another method of oxidation is breaking glucose and glycogen molecules into several fragments and letting these fragments oxidize each other. This means energy yield under *anaerobic* conditions, no external oxygen is imported; the processes are called glycolysis and glycogenolysis, respectively. The energy yield is much less than in aerobic metabolism.

Glucose catabolism (and fat catabolism as well) takes place in sequential biochemical reactions in which intermediary metabolites are produced. The first phase is anaerobic: the 6-carbon compound glucose is broken into two 3-carbon molecular fragments, each of which naturally becomes a 3-carbon compound pyruvic acid molecule. This can become lactic acid if not oxidized, important for muscle fatigue, as discussed in Chapter 3. Some energy is released in this process, described in stoichiometric terms as

$$C_6H_{12}O_6 + 1\,\text{ATP} = 2\,\text{lactic acids} + 3\,\text{ATP} \qquad (8\text{-}5)$$

The second phase starts out to be anaerobic: pyruvic acid is split into carbon dioxide and hydrogen H in a series of self-renewing reactions known as the Krebs cycle, also called the citric acid cycle (see Fig. 8-3). Here, hydrogen atoms are separated in pairs from the intermediary metabolites, the first of which is pyruvic acid. (The removal of hydrogen atoms from the intermediary metabolites is particular to the Krebs cycle and is called dehydrogenation). As oxygen becomes available, hydrogen reacts with it to form water. Now, glucose is completely metabolized and six CO_2 molecules and six H_2O molecules are produced, as in equation (8-4).

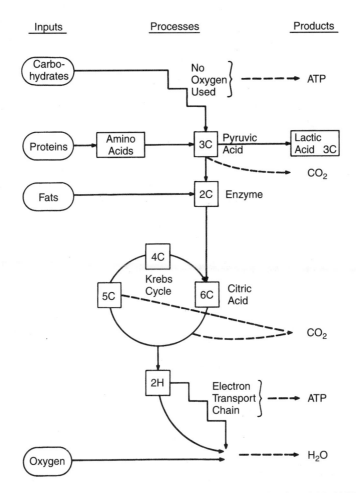

Figure 8-3. Breakdown of foodstuffs (modified from Astrand and Rodahl, 1977), showing the number of carbon atoms in the steps.

Fatty Acids and Amino Acids

For energy release, neutral fat is first split into the 3-carbon metabolite glycerol and the 2-carbon fatty acids. Glycerol is used similarly to glucose, with some energy yield. Fatty acid (and amino acid from protein) is reformed to acetic acid, which enters the Krebs cycle and finally becomes oxidized to carbon dioxide and water. The final energy gain is about 2,340 kcal/mole, more than three times that from glucose.

Under normal nutritional and exertion conditions, fat accounts for most of the stored energy reserves. Glycogen and glucose are the most easily and first used sources of energy at the cell level, particularly in the central nervous system and at muscles.

Energy Storage

Fat provides by far the most stored energy. On average, about 16% of body weight is fat in a young man, increasing to 22% by middle age and more if he "is fat." Young women average about 22% of body weight in form of fat, rising to 34% or more by middle age. Athletes generally have lower percentages, about 15% in men and 20% in women. Assuming 15% of a 60-kg person as a low value results in 9 kg of body fat. Twenty-five percent of a 100-kg person, or 34% of a 75-kg person, means approximately 25 kg of body fat. This amounts to an energy storage in fat of about 85,500 kcal for a skinny lightweight person, and nearly 240,000 kcal for a heavy person.

The energy yield of fat per volume is approximately 2,340 kcal/mole, that is, nearly 3.5 times that of the carbohydrate derivatives glucose and glycogen—but it takes more time to get started.

Glycogen provides much less stored energy. Most of us have about 400 g of glycogen stored near the muscles, about 100 g in the liver, and some in the bloodstream. This means that we have only some 2,200 kcal as energy available from glycogen. About the same amount is present in the form of glucose.

Under normal circumstances, the body does not use protein amino acids for energy since their catabolism usually involves the death of cells, protein being part of the protoplasm of living cells. Catabolism of proteins does occur, however, in fasting, malnutrition, starvation, certain illnesses, and in all-out physical efforts.

Energy for Muscle Work

In muscle, the mitochondria cells store quick-release energy in the molecular compound adenosine triphosphate, ATP. Its phosphate bonds can be broken down easily by hydrolysis, thus providing quick energy for muscle contraction.

$$ATP + H_2O = ADP + energy \text{ (output)} \tag{8-5a}$$

An intracellular carrier of chemical energy, ATP transfers its energy by donation of its high-energy phosphate group to processes that require energy, such as muscle contraction. But the ATP supply is very quickly consumed and needs to be replenished constantly. This is done through creatine phosphate, CP, which transfers a phosphate molecule to adenosine diphosphate, ADP. Energy must be supplied for this reaction:

$$ADP + CP + energy \text{ (input)} = ATP + H_2O \tag{8-5b}$$

The cycle of converting ATP into ADP, releasing energy for muscle action, and reconverting ADP into ATP is anaerobic.

   While ATP provides quick energy for a few seconds, resynthesis of ATP is necessary for continuous operation. This requires energy, which is provided by the breakdown of glucose, glycogen, fat, and protein. The breaking of their complex molecules to simpler ones, ultimately to CO_2 and H_2O, is the ultimate energy source for the ADP-to-ATP reconversion.

   Returning to our earlier analogy, the "human energy machine" is a battery or generator (releasing energy through ATP-to-ADP conversion) recharged or driven by a combustion engine (fueled by glucose, glycogen, and fat).

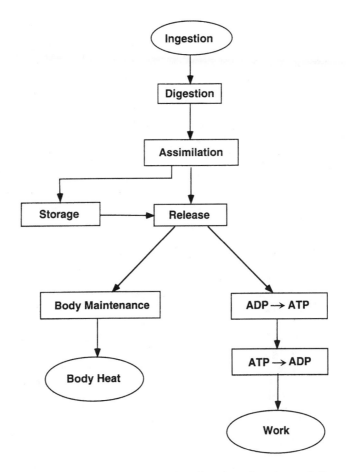

Figure 8-4. Schematic view of the energy flow from ingestion via metabolism to output.

The First Few Seconds

At the very beginning of muscular effort, breaking the phosphate bond of ATP releases quick energy for muscular contraction. However, the contracting muscle consumes its local supply of ATP in about 2 sec.

The First Ten Seconds

The next source of immediate energy is CP. Creatine phosphate transfers a phosphate molecule to the just created molecule of ADP and, thus, turns it back into ATP. (This cycle of converting ATP into ADP and back to ATP does not require the presence of oxygen.) Since we have three to five times more ADP than ATP, there is enough energy available to a muscle for up to 10 sec of vigorous activity.

Longer Than Ten Seconds

After about 10 sec of ATP-ADP-ATP reactions, energy must be supplied to sustain the reformatting of ATP. Now, the energy absorbed from the foodstuffs comes into play: glucose is broken down, releasing energy for the recreation of ATP.

Given that only a few seconds have elapsed since the activity started, there simply has not been enough time to use oxygen in the energy conversion process. Thus, while releasing energy, the breakdown of glucose (generating carbon dioxide and water) is not complete: other metabolic by-products are also generated, particularly lactic acid. (If this metabolic by-product is not resynthesized within about a minute in the presence of oxygen, the muscles simply cannot continue to do work.)

This anaerobic energy release relies primarily on the breakdown of glucose although some glycogen is also involved. This is why glucose is called the primary, most easily accessible, and most metabolized energy carrier for the body.

Minutes and Longer

If the physical activity has to continue, it must be performed at a level at which oxygen is sufficiently available to maintain the energy conversion processes. Hence, the energy generated by a quick burst of maximal effort cannot be maintained at this level for extended periods of time.

In enduring work, the energy demanded from the muscles is so low that the oxygen supply at the cell level (mitochondria) allows aerobic energy conversion, meaning that sufficient oxygen is available to maintain the energy processes. Without oxygen, 1 molecule of glucose yields 2 molecules of ATP. With oxygen, the glucose energy yield is 39 molecules of ATP. Even richer in energy is the fat molecule palmitate, which yields about 130 molecules of ATP.

Aerobic and Anaerobic Work

Because energy yield is so much more efficient under aerobic conditions where no metabolic by-products are generated, a fairly high energy expenditure can be sustained as long as ATP is replaced as quickly as it is used up. The energy yield of glucose under aerobic conditions releases energy in the amount of 690 kcal/mole. (The conversion is complete, and no lactic acid develops.) A more complex conversion takes place in the utilization of fats (glycerol and fatty acids), which are converted to intermediary metabolites and enter the Krebs cycle. Their final energy yield is approximately 2,340 kcal/mole, more than three times that from glucose.

Still, if very heavy expenditure is required over long periods of time, such as in a marathon run, the interacting metabolic system and the oxygen-supplying circulatory system might become overtaxed. The runner who "hits the wall" has most likely used up the body's glycogen supply and has also gone into "oxygen debt."

In our regular activities, however, the energy output is regulated to agree with the body's abilities to develop energy under sufficient supply of oxygen. If needed, we simply take a break and, while resting, accumulated metabolic by-products are resynthesized and the metabolic, circulatory, and respiratory systems are returned to their normal states.

Overall, sustained work is aerobic, but most of the single intermediate steps in the metabolic reactions are in fact anaerobic. Glucose breakdown is first anaerobic, followed by an aerobic phase, but oxygen is required for complete metabolism. Glycogen stores near muscles are depleted much more quickly when the muscles must work anaerobically than when they are able to work aerobically. The combustion of the fat derivatives is strictly aerobic.

Usually, at rest and during moderate work, the oxygen supply is sufficient and, hence, the energy metabolism is essentially aerobic. This leads to high ATP and low ADP concentrations. To meet intermediate work (energy) demands, the breakdown of glucose speeds up. At a critical intensity, the oxygen-transporting system cannot provide enough oxygen to the cells and pyruvate is transformed into lactic acid instead of going through the Krebs cycle. During continued intermediate work, the lactate developed may be reconverted to glycogen in the liver, and perhaps even at the muscle (Astrand and Rodahl, 1986) if aerobic conditions exist. With high work intensity, increasingly more parts of the metabolic processes are anaerobic; lactic acid buildup increases, which may eventually require cessation of the muscular work.

Looking more minutely at the processes of energy liberation during the first 45 to 90 sec of exercise, we find that, when the work intensity is between 50% and 100% of the maximum possible, some anaerobic energy metabolism takes place, while the oxygen supply is not yet adjusted to meet the demand.

Table 8-2 shows schematically the contributions of aerobic and anaerobic energy liberation during short and long maximal efforts. Of course, the intensity magnitude of the effort is higher in a short burst of output than in sustained exertion.

During maximal work of short duration (up to 1 min or so), the energy available depends on ATP splitting. During prolonged heavy exercise (longer than 1 hr), the maximal work output depends on the oxidation of fatty acids and glycogen. The most complex process in terms of fuel conversion is during hard work that lasts from about 1 to 10 min. At the start of such rigorous work, utilization of ATP and PC is predominant. Then anaerobic conversion of glucose to lactate takes over increasingly.

During the final phase, the oxidation of glucose and, eventually, of fatty acids predominates.

Table 8-2. Contributions of anaerobic and aerobic energy liberation during maximal efforts of various durations (schematically from Astrand and Rodahl 1977).

Energy released	Duration of the greatest possible effort				
	10 s	1 min to	10 min	1 hr to	2 hr
By anaerobic processes					
in kJ	100	170	150	80	65
in %	85	65 to 70	10 to 15	2	1
By aerobic processes					
in kJ	20	80	1,000	5,500	10,000
in %	15	30 to 35	85 to 90	98	99
Total in kJ	120	250	1,150	5,580	10,065
Primary energy source	ATP splitting	CP	Glucose glycolysis	Glucose and fatty acids Krebs cycle	
Process	Anaerobic	Mixed		Aerobic	

Energy Use and Body Weight

Equation (8-1) describes the balance between energy input, energy output, and energy storage. If the input exceeds the output, storage (body weight) is increased; conversely, a weight decrease takes place if the input is less than the output. About 7,000 to 8,000 kcal make a difference of 1 kg in body weight.

The body tries to maintain a given energy storage. This means that a person's body weight—which, in addition to water, consists mostly of the weight of bones, tissues, and fat as stored energy—is kept at this level unless the "set point" is altered by radical changes in health, in energy expenditures (such as continued vigorous exercising or reduction of physical activities), and in nutritional habits (such as severe overeating or starvation). As long as the body maintains the set point, body weight remains rather constant. Even if food intake is reduced, the body tries to extract enough energy from the reduced food intake to maintain the old body weight. A continued starvation diet usually lowers storage and hence body weight. Returning to the previous eating habits after having achieved the desired lowered body weight allows the body to re-attain its previous weight unless the set point has been lowered.

Summary (of Section 1)

The metabolic breakdown of foodstuffs releases energy that is partially transmitted as useful mechanical energy to an outside object—"work"—but is mostly transformed into heat.

Overall, the human metabolic process liberates energy (exergonic), but reactions within this process require energy (are endergonic).

The primary energy suppliers are carbohydrates and fats. Protein provides enzymes as catalysts in the digestion, and it is used for tissue rebuilding but, normally, not for energy.

The oxidation of foodstuffs provides the overall energy source for the metabolic processes, while the breakdown of ATP into ADP is the immediate energy source.

The breakdown of glucose is in aerobic as well as anaerobic steps, with the latter often resulting in the buildup of lactic acid—a metabolite that can force cessation of the effort if it is not resynthesized. Glucose provides most of the energy for maximal efforts lasting less than 1 hr.

The breakdown of fatty acids is aerobic that is, needs oxygen. This provides much of the energy for heavy physical work lasting hours.

SECTION 2. ASSESSMENT OF ENERGY EXPENDITURES

Overview

To match a person's work capacity with the job requirements, one needs to know the individual's energetic capacity and how much a given job demands of this capacity. This section primarily addresses the first topic; the next section deals with job demands.

Among the currently used procedures to assess internal metabolic capacities, four techniques predominate:

Diet and weight observation
Direct calorimetry
Indirect calorimetry
Subjective rating of perceived strain

Introduction

Individuals vary greatly in their ability to do physical work, with body size and conditioning, age and health, attitude and environment as major factors; see Figure 8-5.

The following text discusses several well-established techniques to measure human energy use and capabilities. However, insight and caution are in order here so that the results of these techniques can be properly understood and applied.

Several challenges make it laborious to assess human energy processes:

- Challenge 1. Human energy balance must be observed over a sufficiently long time period so that delays and advances in energy intake, conversion, and output "average out." In equations (8-2) and (8-3), energy storage was (for convenience) excluded. However, imbalances between energy intake and output of hundreds of kilocalories over a day are quite common. Most of the energy store in a healthy adult is in the form of adipose tissue of the magnitude of about 100,000 kcal. A substantial change in storage should result in a change in body weight, which is easily observable. (Many data on energy consumption are given in relation to body weight.) However, the

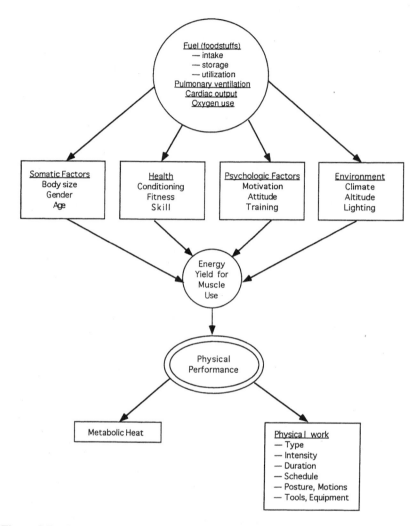

Figure 8-5. Determiners of individual physical work capacity (with permission from Kroemer, K.H.E., Kroemer, H.B., and Kroemer-Elbert, K.E. (1994). *Ergonomics: How to Design for Ease and Efficiency.* Englewood Cliffs, NJ: Prentice-Hall. All rights reserved.)

seemingly simple relationship between weight and energy store is disturbed by water, which is a large and rather labile component of body weight but contributes nothing to the energy stores (Garrow, 1980).

- Challenge 2. Assessment of the amount of total body fat or, conversely, of lean body mass is usually done by measuring skinfold thickness and inserting the results into various equations that should yield values for body fat or lean mass (Roebuck, Kroemer, and Thompson, 1975). Altogether, this is not very accurate.

- Challenge 3. The procedures and conditions of measuring the energy exchanges may influence the measured results: putting on and carrying the measuring apparatus, hindrance of motions and breathing, and simply being observed may lead to energetic and circulatory processes different from those under normal circumstances.
- Challenge 4. Inaccuracies in the measurements and inadequacies in the experimental procedure may degrade the results. As discussed in more detail later, energy output measurements should be taken during the "steady state" of work, after the functions have reached stability, and well before a rest period. This may be rather difficult to do, for example, if the time available for the measurement is very short or if work intensity or the environment change, preventing a steady state from developing. In fact, there may be no steady state in short-term efforts or in efforts near or beyond a person's capacity.
- Challenge 5. For data, we need a baseline, or reference. We would like to start with the basal metabolism (or heart rate) and observe the increase from this value. However, a subject's true, reliable basal condition is quite difficult to achieve; hence, observed differences from that basal value are subject to unaccounted variations in the baseline and, of course, in the values measured under stress. This is of particular concern for light efforts: assume that the measuring accuracy for energy at both rest and work is in the neighborhood of 5%, and assume a "true" energy expenditure increase at work of 10% over rest. In this case, one might actually observe differences in expenditures between 0% and 20%; the first value is obviously nonsense, and the second is double the "truth." (This may, in fact, explain some of the percentwise enormous variations found in different texts concerning the energy requirements of light work.)
- Challenge 6. In experimental design, we need to distinguish clearly between the independent variable, the effects of which we want to study in the experiment; and the dependent variable, which provides that information. For example, we may wish to measure the individual capacity of a person to perform standardized work of known intensities. The dependent measure would be the oxygen consumption or heart rate, or another strain indicator of the body. In the laboratory, standardized loads (for example, using a bicycle ergometer or treadmill) allow judgments about an individual's capacity related to such workloads. This information "standardizes" the subject (with respect to the reactions to such workloads), who then can be used to assess how much of a strain it is to do certain unknown work, for example, in industry, or on a construction site, or on the moon.

Simple and well-controlled laboratory conditions do not exist in the real world. Here, one often has to assess the "heaviness" of work through the reactions of subjects that have not been standardized. Many of the data describing the energy expenditures, or heart rates, associated with certain jobs or professions or occupations (see Sec. 3) must be applied cautiously since there is little or no assurance that the field data on energy consumption or heart rate were indeed measured on "standard subjects." The wide diversity in existing data may be due to variability in the subjects employed and to variations in their tasks.

Diet and Weight Observation

In terms of energy, what is inputted into the body as nutrients, solid or fluid, must be outputted either in terms of energy internally consumed to maintain the body (finally converted to heat) or as external work performed. This "energy balance" assumes that there is neither energy storage (when the person gets heavier, particularly fatter) nor use of stored energies (when the person gets thinner). Assessment of internal energetic processes through diet and weight observation provides reliable information about the long-term energy exchange but requires rather lengthy observation periods (weeks or more). Strict, complete control over food and drink intake, water balances, and energy output is feasible in the laboratory but rather difficult in everyday applications.

Direct Calorimetry

Since energy taken in by the body is finally transformed into heat (if no external work is done), one can enclose the human body with an energy-tight chamber that allows measurement of all heat generated and transmitted by conduction, convection, radiation, and evaporation. There are several requirements for this procedure: the room must be small, which reduces the capability to perform work; air, which carries energy, must be supplied to, and taken from, the chamber; since the setup is not massless, energy is exchanged with, and stored in, equipment, walls, and so forth. Hence, direct calorimetry must be performed in special laboratories and is not of much practical interest. Most current calorimetric techniques rely on indirect calorimetry.

Indirect Calorimetry

Assessment by Oxygen Consumption

While work is performed, the oxygen consumption (along with CO_2 release) is a measure of metabolic energy production, which can be assessed with a variety of instruments. (Brief descriptions of the techniques to measure oxygen consumption are provided in Appendix A to this chapter.) All the techniques rely on the principle that the difference in O_2 (and CO_2) content between the exhaled and inhaled air indicates the oxygen absorbed (or carbon dioxide released) in the lungs. Given sufficiently long observation periods (at least, 5 min), this is a reliable assessment of the metabolic processes. Assuming an overall "average" caloric value of oxygen of 5 kcal/L O_2, we can calculate the energy conversion occurring in the body from the volume of oxygen consumed.

A more exact assessment of the nutrients actually metabolized can be made by using the respiratory exchange quotient, RQ, which compares the carbon dioxide expired to the oxygen consumed. One gram of carbohydrate needs 0.83 L of oxygen to be metabolized and releases the same volume of carbon dioxide [see equation (8-4)]. Hence, the RQ is 1 (unit). The energy released is 18 kJ/g, 21.2 kJ/L O_2 or 5.05 kcal/L O_2. Table 8-3 shows these relationships also for fat and protein conversion.

Most medical and physiological assessments of human energetic capabilities currently utilize various techniques of measuring oxygen consumption. To compare one person's capacities with another's, we rely on standardized tests with normalized external work, using mostly bicycle ergometers, treadmills, or steps. Selection of this equipment has often been based on availability and ease of use as much as on theoretical considerations.

Bicycles, treadmills, and steps exercise certain body parts and body functions. Bicycling requires predominantly use of the leg muscles. Since the legs, both in their mass and their musculature, constitute rather large components of the human body, their extensive exercising in a bicycle test also strains pulmonary, circulatory, and metabolic functions of the body. However, a person who is particularly strong in the upper body but not well trained in the use of legs would show different strain reactions in bicycle ergonometry than, say, a bicycle racer.

The treadmill also strains primarily lower body capabilities but, in contrast to bicycling, the whole body weight must be supported and propelled by the feet. If the treadmill is inclined, the body must also be lifted. Hence, this test strains the body in a somewhat more complete manner than bicycling but still leaves trunk and arm capabilities out of consideration. Furthermore, it requires somewhat more bulky equipment than a stationary ergometer bicycle.

Another method of standardizing stress conditions is to have the subject step onto a platform, step down, step up, and so forth, for the duration of the test. This step test technique requires the simplest possible equipment and strains body functions in a fashion somewhat similar to running on a treadmill, however, by making the subject elevate the total body weight instead of moving it forward. Again, muscular capabilities of the upper body are not tested at all, but a heavier person with shorter legs than another subject would show larger energy consumption.

Obviously, these three techniques differ in that they involve different body masses and muscles and, hence, strain local (muscular) or central (pulmonary, circulatory, metabolic, etc.) capabilities in somewhat different ways. Improvements on the equipment providing the external stress have been proposed in many different ways. Two examples of exercise equipment that strains the body more uniformly

Table 8-3. Oxygen needed, RQ, and energy released in nutrient metabolism.

	O_2 Consumed $(L\ g^{-1})$	$RQ = \dfrac{Vol\ CO_2}{Vol\ O_2}$	$kJ\ g^{-1}$	$kJ\ L^{-1}O_2$	$kcal\ L^{-1}O_2$
Carbohydrate	0.83	1.00	18	21.2	5.05
Fat	2.02	0.71	40	19.7	4.69
Protein	0.79	0.80	19	18.9	4.49
Average*	NA	NA	NA	21	5

*Assuming the construct of a "normal" adult on a "normal" diet doing "normal" work.

are a ladder-mill on which one climbs, using both arms and legs, and an ergometer bicycle with simultaneous cranking or pushing and pulling with the hands.

Whichever method is used to develop the external stress, the reactions of the individual in terms of oxygen consumption and, hence, energy metabolism may be collected and compared to the data gained from other subjects.

Assessment by Heart Rate

There is close interaction between the circulatory and metabolic processes. For proper functioning, nutrients and oxygen must be brought to the muscle or other metabolizing organs and metabolic by-products removed from it. Therefore, heart rate as a primary indicator of circulatory functions, and oxygen consumption, representing the metabolic conversion taking place in the body, have a linear and reliable relationship in the range between light and heavy work, shown in Figure 8-6. This relationship may change within one person with training, and it differs from one individual to another. Given this relationship, heart rate measurements can simply be substituted for measurement of metabolic processes, particularly for O_2 assessment. This is a very attractive shortcut since heart rate measurements can be performed easily.

The simplest way to count the heart rate (HR) is to palpate an artery, often in the wrist or perhaps the neck, or to listen to the sound of the beating heart. All the

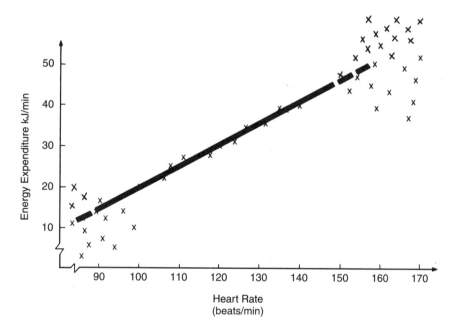

Figure 8-6. Relationships between oxygen uptake (expressed as energy expenditure) and heart rate.

measurer needs to do is to count the number of heartbeats over a given period of time (such as 15 sec) and, from this, calculate an average heart rate per minute. More refined techniques utilize various plethysmographic methods, which rely on tissue deformations due to changes in filling the embedded blood vessels. These methods range from mechanically measuring the change in volume of tissues in a finger, for example, to using photoelectric techniques that react to changes in transmissibility of light depending on the blood filling of an ear lobule, for instance. Other techniques rely on electric signals associated with the pumping actions of the heart, sensed by electrodes placed on the chest.

These techniques have limited reliability in assessing metabolic processes primarily by the intra- and interindividual relationships between circulatory and metabolic functions. Statistically speaking, the regression line (shown in Fig. 8-6) relating heart rate to oxygen uptake (energy production) is different in slope and intersect from person to person and from task to task. In addition, the scatter of the data around the regression line, indicated by the coefficient of correlation, is also variable. The correlation is low at light loads, where the heart rate is barely elevated and circulatory functions can be influenced easily by psychological events (excitement, fear, etc.) which may be unrelated to the task. With very heavy work, the O_2-HR relation may also fall apart, for example, when cardiovascular capacities are exhausted before metabolic or muscular limits are reached. The presence of heat load also influences the O_2-HR relationship.

Heart rate measurement has a major advantage over oxygen consumption as an indicator of metabolic processes: heart rate responds more quickly to changes in work demands and, hence, indicates more readily quick changes in body functions due to changes in work requirements.

Assessment by Subjective Rating of Perceived Effort

Human beings are able to perceive the strain generated in their bodies by given tasks and to make absolute and relative judgments about perceived efforts: we can assess and rate the relationships between physical stimulus (the work performed) and its perceived sensations (Pandolph, 1983). This correlation between the psychologically perceived intensity of physically described stimuli has been used probably as long as people have existed to express their preferences for one type of work over another. In the 1970s, Borg developed formal techniques for rating the perceived exertion (RPE) associated with different kinds of efforts. For example, perceived effort can be assessed using a nominal scale ranging from "very light" to "hard." Such a verbally anchored scale can be used to "measure" the strain subjectively perceived by a subject performing standardized work, allowing a relative assessment of that person's capability to perform stressful work. (Examples of Borg scales are given in Appendix B at the end of this chapter.)

Summary (of Section 2)

Among the techniques for assessing metabolic processes, direct calorimetry and observation of diet and weight are feasible under well-controlled laboratory conditions. More practical are the measurement of oxygen uptake and, relatedly, of heart rate. These measurements can be performed over relatively short periods of time with commercially available equipment. (Techniques are discussed in detail, e.g., by Kinney, 1980; Mellerowicz and Smodlaka, 1981; Stegemann, 1984; Eastman Kodak, 1983, 1986; Webb 1985.) Their primary utility lies in the assessment of a given person's metabolic/circulatory capabilities in response to standardized work, often measured using ergometers of the bicycle or treadmill types. Another common technique is Borg's rating of subjectively perceived effort in the performance of certain work.

SECTION 3. ENERGY REQUIREMENTS AT WORK

Overview

This section discusses the demands made on the body metabolism by different activities.

Procedures for Categorizing Metabolic Requirements

Assuming that, without energy storage, all metabolically developed energy is transformed into either externally performed work or, finally, heat, we customarily assess the energy requirements imposed on the body within categorized activity levels.

Basal Metabolism

A minimal amount of energy is needed to keep the body functioning, even if no activities are performed at all. This basic metabolism is measured under strict conditions, usually including fasting for 12 hr, protein intake restriction for at least two days, with complete physical rest in a neutral ambient temperature. Under these conditions, the basal metabolic values depend primarily on age, gender, height, and weight, with the last two variables occasionally replaced by body surface area. Altogether, there is relatively little interindividual variation; hence, a commonly accepted figure is 1 kcal (4.2 kJ)/kg/hr (or 4.9 kJ/min for a person of 70 kg).

Resting Metabolism

The highly controlled conditions under which basal metabolism is measured (mostly for medical purposes) are rather difficult to accomplish in practical applications. Metabolism is usually measured before the working day, with the subject as much at rest as possible. Depending on the given conditions, resting metabolism is 10% to 15% higher than basal metabolism.

Work Metabolism

The increase in metabolism from resting to working is called work metabolism. This increase above resting level represents the amount of energy needed to perform the work.

At the start of physical work, oxygen uptake follows demand sluggishly. As Figure 8-7 shows, after a slow onset, oxygen intake rises rapidly and then slowly approaches the level at which the oxygen requirements of the body are met. During the first minutes of physical work, there is a discrepancy between oxygen demand and available oxygen. (During this time, the energy yield is largely anaerobic.) This oxygen deficit must be repaid at some time, usually during rest after work. The amount of this deficit depends on the kind of work performed and on the individual.

After stopping the work, the oxygen demand falls again to a resting level, quickly at first and then leveling off. During this time, the depleted ATP stores are refilled, but elevated tissue temperature and epinephrine concentration, increased cardiac and respiratory functions, and other phenomena require (as Astrand and Rodahl, 1986, put it) that "the body pay 100% interest on the oxygen borrowed from the anaerobic bank": the oxygen debt repaid is approximately twice as large as the oxygen deficit incurred. Of course, given the close interaction between the circulatory and the metabolic systems, heart rate reacts similarly; but, as it increases faster at the start of work than oxygen uptake, it also falls back more quickly to the resting level.

An explanation for the sluggish responses of oxygen intake to the start of external physical work lies in the time it takes to increase the flow of oxygen-rich blood to the internal consumers. Some oxygen is stored in the muscles bound to myoglobin and in the blood profusing the muscles. At the beginning of muscular labor, this stored oxygen does not suffice, and anaerobic processes take place to release energy for the muscular work.

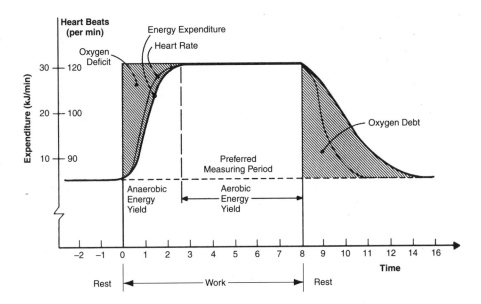

Figure 8-7. Schematic illustration of energy liberation, energy expenditure, and heart rate at steady-state work.

This results in the production of lactic acid. Insufficient blood flow also leads to excessive buildup of potassium (Kahn and Monod, 1989). Lactic acid and potassium accumulation are believed to be the primary reasons for "muscle fatigue" (discussed in some detail later) forcing the stoppage of muscle work.

If the workload does not exceed approximately 50% of the worker's maximal oxygen uptake, then oxygen uptake, heart rate, and cardiac output can finally achieve the required supply level and can stay on this level. This condition of stabilized functions at work is called *steady state*. Obviously, a well-trained person can attain this equilibrium between demands and supply even at a relative high workload, while an untrained person would be unable to attain a steady state at this requirement level but could be in equilibrium at lower demand.

If the energetic work demands exceed about half the person's maximal O_2 uptake capacity, anaerobic energy-yielding metabolic processes play increasing roles. In particular, this results in high lactic acid production. The length of time during which a person performs this work depends on the individual's motivation and will to overcome the feeling of "fatigue" (see below), which usually coincides with depletion of glycogen deposits in the working muscles, drop in blood glucose, and increase in blood lactate. However, the physiological processes involved are not fully understood, and highly motivated subjects may maintain work that requires very high oxygen uptake longer than subjects who feel that they must stop after just a few minutes of effort.

When vigorous exercise brings about a continuously growing oxygen deficit and an increase in lactate content of the blood because of anaerobic metabolic processes, a balance between demands and supply cannot be achieved; no steady state exists, and the work requirements exceed capacity levels; this is sketched in Figure 8-8. Impending exhaustion can be counteracted by the insertion of rest periods. Given the same ratio of total resting time to total working time, many short rest periods have more recovery value than a few long rest periods.

It is apparent that calorimetric or heart rate measurements should not be taken during the rising phases of oxygen intake or heart rate but that these measurements should be performed during the steady-state period, when the physiological functions have achieved equilibrium. Measurements during the first few minutes are not indicative of the demand-and-supply functions between work and body. Likewise, measurement after the cessation of the work is not a direct indicator of the severity of the preceding work demands and of their physiological responses; under certain highly controlled conditions, however, counting the recovery pulse has been suggested. Here, the heart rate is monitored from the exact moment at which the work was stopped throughout predetermined intervals of the recovery period.

Techniques for Estimating Energy Requirements

Using Tables

There are various ways to describe work or exercise activities: by the amount of work to be done with the arms or the legs, by the involvement of the trunk, by the total amount of external work being done, each over the periods of time during which these activities occur. Classic such data for many jobs or occupations

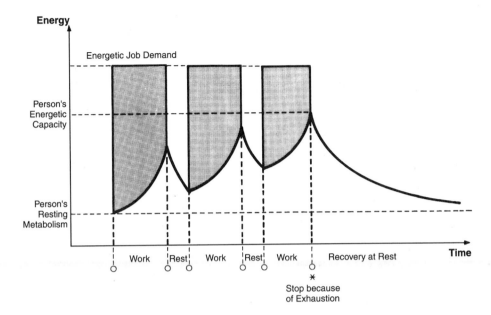

Figure 8-8. Metabolic reactions to the attempt to do work that exceeds one's capacity, even with interspersed rest periods.

were compiled by Spitzer and Hettinger in 1958 and Durnin and Passmore in 1967. Excerpts from information on energy expenditures in various body motions and job and sports activities are compiled in Tables 8-4 through 8-6. These tables should be used with caution, however, as explained at the beginning of this section. Also, note that Table 8-4 refers to work metabolism. Table 8-5 contains information on total energy cost per day, and Table 8-4 relates to kilograms of body weight.

Calculating

Instead of using general tables that contain the energy expenditures of, for example, lumbermen, carpenters, housewives, and secretaries, we can compose the total energetic cost of given work activities by adding up the energetic cost of the work elements that, taken together, make up this activity. This "analytic" approach largely avoids the problems associated with the fact that the data in the "synthetic" tables (such as Table 8-5) found in older publications reflect measurements taken when many activities were quite different from what they are today: these days, housewives seldom wash and wring by hand, secretaries do not pound mechanical typewriters, lumbermen no longer cut down trees with handsaw and ax, and farmers do not walk behind plows pulled by horses or mules. Furthermore, these data may be "averaged" over unknown subjects or attained with only a few subjects and, hence, may be inaccurate or unreliable.

Table 8-4. Energy consumption (to be added to basal metabolism) at various activities (adapted from Astrand and Rodahl 1977, Guyton 1979, Rohmert and Rutenfranz 1983, Stegemann 1984).

	kJ min^{-1}
Lying, Sitting, Standing	
Resting while lying	0.2
Resting while sitting	0.4
Sitting with light work	2.5
Standing still and relaxed	2.0
Standing with light work	4.0
Walking without load	
on smooth horizontal surface at 2 km hr^{-1}	7.6
on smooth horizontal surface at 3 km hr^{-1}	10.8
on smooth horizontal surface at 4 km hr^{-1}	14.1
on smooth horizontal surface at 5 km hr^{-1}	18.0
on smooth horizontal surface at 6 km hr^{-1}	23.9
on smooth horizontal surface at 7 km hr^{-1}	31.9
on country road at 4 km hr^{-1}	14.2
on grass at 4 km hr^{-1}	14.9
in pine forest, on smooth natural ground at 4 km hr^{-1}	18 to 20
on plowed heavy soil at 4 km hr^{-1}	28.4
Walking and carrying on smooth solid horizontal ground	
1 kg load on back at 4 km hr^{-1}	15.1
30 kg load on back at 4 km hr^{-1}	23.4
50 kg load on back at 4 km hr^{-1}	31.0
100 kg load on back at 3 km hr^{-1}	63.0
Walking downhill on smooth solid ground at 5 km hr^{-1}	
5° decline	8.1
10° decline	9.9
20° decline	13.1
30° decline	17.1
Walking uphill on smooth solid ground at 2.5 km hr^{-1}	
10° incline (gaining height at 7.2 m min^{-1})	
no load	20.6
20 kg on back	25.6
50 kg on back	38.6
16° incline (gaining height at 12 m min^{-1})	
no load	34.9
20 kg on back	44.1
50 kg on back	67.2
25° incline (gaining height at 19.5 m min^{-1})	
no load	55.9
20 kg on back	72.2
50 kg on back	113.8
Climbing stairs 30.5° incline, steps 17.2 cm high	
100 steps per minute (gaining 17.2 m min^{-1}), no load	57.5
Climbing ladder 70° incline, rungs 17 cm apart	
(gaining 11.2 m min^{-1}), no load	33.6

Note: While Rohmert and Rutenfranz (1983) claim that intra- and inter-individual differences in energy consumption are within ± 10% for the same activity, a comparison of data presented in various texts shows a much higher percentage in variation, particularly at activity levels requiring little energy.

Table 8-5. Total energy cost per day in various jobs and professions (adapted from Astrand and Rodahl 1977). Note that the physical job demands may be different today from what they were decades ago.

Occupation	Energy expenditure, kcal/day		
	Minimum	**Mean**	**Maximum**
Men			
Laboratory technicians	2240	2840	3820
Elderly industrial workers	2180	2840	3710
University students	2270	2930	4410
Construction workers	2440	3000	3730
Steel workers	2600	3280	3960
Elderly peasants (Swiss)	2210	3530	5000
Farmers	2450	3550	4670
Coal miners	2970	3660	4560
Forestry workers	2860	3670	4600
Women			
Elderly housewives	1490	1990	2410
Middle-aged housewives	1760	2090	2320
Laboratory technicians	1340	2130	2540
University students	2090	2290	2500
Factory workers	1970	2320	2980
Elderly peasants (Swiss)	2200	2890	3860

Table 8-6. Total energy consumption per kg body weight at various sports activities (adapted from Stegemann 1984).

Sports Activity	Energy Consumption $kJ\ kg^{-1}\ hr^{-1}$
Cross-country skiing, 9 km/hr	38
Jogging, 9 km/hr	40
Ice skating, 21 km/hr	41
Swimming (breaststroke), 3 km/hr	45
Running, 12 km/hr	45
Running, 15 km/hr	51
Wrestling	52
Playing badminton	53
Running, 17 km/hr	60
Bicycling, 43 km/hr	66
Cross-country skiing, 15 km/hr	80

If we know the time spent in a given activity element and its metabolic cost per time unit, we can simply calculate the energy requirements of this element by multiplying its unit metabolic cost with its duration time. Take a simple case: For a person resting (sleeping) 8 hr/day, at an energetic cost of approximately 5.1 kJ/min, the total energy cost is 2,448 kJ (5.1 kJ/min × 60 min/hr × 8 hr). If the person then does 6 hr of light work while sitting, at 7.4 kJ/min, this adds another 2,664 kJ to the energy expenditure. With an additional 6 hr of light work done standing, at 8.9 kJ/min and further with 4 hr of walking at 11.0 kJ/min, the total expenditure during the full 24-hr day would come to 10,956 kJ (approximately 2,610 Cal). Similar approaches based on industrial engineering practices, related mostly to job analysis (Bernard and Joseph, 1994), can be successful and convenient.

As this example shows, the energetic requirements of given activities—per hour, day, week, or year—can be computed from tables of metabolic requirements of certain job elements. In using these, care should be taken to check whether or not they include the basal or resting rates. For example, Table 8-4 does not contain the basic value, while Tables 8-5 and 8-6 do. In developed countries, daily expenditures range from about 6,000 to 20,000 kJ/day, with observed median values of about 10,000 kJ (2,400 Cal) for women and about 14,000 kJ (3,300 Cal) for men.

Light or Heavy Jobs?

Energy requirements allow us to judge whether a job is (energetically) easy or hard. Given the largely linear relationship between heart rate and energy uptake, we can often use heart rate to establish the "heaviness" of work. Of course, such descriptive labels as "light" or "easy" or "heavy" reflect judgments that rely very much on the current socioeconomic concept of what is permissible, acceptable, comfortable, easy, or hard. Depending on the circumstances, we find a diversity of opinions about how physically demanding a given job is. One such rating of job severity, in terms of energetic or circulatory demands, is presented in Table 8-7.

Table 8-7. Classification of light to heavy work (performed over a whole work shift) according to energy expenditure and heart rate.

Classification	Total energy expenditure		Heart rate in beats/min
	in kJ/min^{-1}	in kcal/min^{-1}	
Light work	10	2.5	90 or less
Medium work	20	5	100
Heavy work	30	7.5	120
Very heavy work	40	10	140
Extremely heavy work	50	12.5	160 or more

This is a unisex table; most men would find the work lighter, and most women experience the effort as heavier than it is labeled.

Light work is associated with rather small energy expenditure (about 10 kJ/min, including the basal rate) and accompanied by a heart rate of approximately 90 beats per minute (bpm). In this type of work, the energy needs of the working muscles are covered by the oxygen available in the blood and by glycogen at the muscle. Lactic acid does not build up. At *medium work,* with about 20 kJ and 100 bpm, the oxygen requirement at the working muscles is still covered, and lactic acid developed initially is resynthesized to glycogen during the activity. In *heavy work,* with about 30 kJ and 120 bpm, the oxygen required is still supplied if the person is physically capable to do such work and specifically trained in this job. However, the lactic acid concentration incurred during the initial minutes of the work is not reduced but remains until the end of the work period, to be brought back to normal levels after cessation of the work.

With light, medium, and even heavy work, metabolic and other physiological functions can attain a steady-state condition throughout the work period, provided the worker is capable and trained. This is not the case with *very heavy work,* where energy expenditures are in the neighborhood of 40 kJ and heart rate is around 140 bpm. Here, the original oxygen deficit increases throughout the duration of work, making intermittent rest periods necessary or even forcing the person to stop this work completely. At even higher energy expenditures, such as 50 kJ/min, associated with heart rates of 160 bpm or higher, lactic acid concentration in the blood and oxygen deficit are of such magnitudes that frequent rest periods are needed and even highly trained and capable persons may be unable to perform this job through a full work shift.

The American Heart Association proposed the MET as the measuring unit for individual work capacity or the metabolic requirement of a task. One MET is defined as the rate of energy expenditure requiring an oxygen consumption of 0.0035 L O_2/min/kg of body weight, which is close to the basal metabolic rate when sitting (Erb, Fletcher, and Sheffield, 1979). This may provide a useful general scale for measuring physical activities; the authors assume that "the average 40-year-old male" in the United States can function at 10 METs at his maximum aerobic capacity.

Another rule of thumb for assessing the magnitude of effort that is energetically permissible has been provided by Astrand and Rodahl (1977, p. 462), who put the limit at 30% to 40% of an individual's maximal oxygen uptake for 8 hr. Exhaustion is likely to occur when more than 50% of maximal activity is required by the job.

As discussed earlier, the oxygen supply to the body can rise from about 0.2 L/min^{-1} at rest to 6 L/min^{-1} (in top athletes) with very hard work. This increase to about 30 times the oxygen demand at rest during work is accompanied by a redirection of the blood supply to the organs in particular need, predominantly to the skeletal muscles that perform physical work. Of the oxygen available at rest, about 20% is usually consumed by muscles, 10% for digestion, 5% by the heart, and around 5% by the brain. At medium work, the muscles will consume about 70% of the oxygen available, while the digestive tract receives only about 5%. The heart, working harder to supply blood, still receives about 5% of the oxygen available, while the brain still consumes about 5%.

Fatigue

As mentioned before, fatigue is an operational description of an individual's temporary state of reduced ability to continue muscular efforts or physical work in general. This phenomenon is best researched for maintained static (isometric) muscle contraction. As described in Chapter 3, when an effort exceeds about 15% of MVC (maximal voluntary contraction), blood flow through the muscle is reduced and it is, in fact, cut off in a maximal effort in spite of a reflex increase in systolic blood pressure. Insufficient blood flow brings about an accumulation of potassium ions in the extracellular fluid, along with depletion of extracellular sodium. Combined with intracellular accumulation of phosphate (from the degradation of ATP), these biochemical events perturb the coupling between nervous excitation and muscle fiber contraction. This decoupling between central nervous system (CNS) control and muscle action signals the onset of fatigue. Depletion of ATP or creatine phosphate as energy carriers and the accumulation of lactate, so far believed to be the reasons for fatigue, do occur but appear not to be the primary reasons. Also, the increase in positive hydrogen ions resulting from anaerobic metabolism causes a drop in intramuscular pH, which then inhibits enzymatic reactions, notably those in the ATP breakdown (Kahn and Monod, 1989).

If work demands approach half a person's maximal O_2 uptake capacity, anaerobic energy-yielding metabolic processes play increasing roles. This results in accumulations of potassium and lactic acid, which are believed to be the primary reasons for the stoppage of muscular work. The length of time during which a person continues to perform this work depends on the individual's motivation to overcome the feelings of fatigue or exhaustion that usually coincide with depletion of glycogen deposits in the working muscles, drop in blood glucose, and increase in blood lactate. The processes involved are not fully understood, however, and highly motivated persons may maintain work that requires very high oxygen uptake for long durations while others stop after just a short period of effort.

Fatigue can be avoided by reducing or stopping the effort. During a rest period, accrued metabolic by-products can be metabolized, the respiratory and circulatory system can recover, and the will to continue can be restored. As shown in Figure 8-6, many short rests have more value in avoiding or overcoming fatigue than fewer longer stops. Applied to heavy shift work, for example (Kroemer, Kroemer, and Kroemer-Elbert, 1994), this means that the manager should encourage workers to take frequent but short breaks throughout the shift rather than trying "to push through and go home early"—and exhausted.

Overall Changes in Body Functions in Response to Workloads

As stated at the beginning of this chapter, there is close interdependency among respiration, circulation, and metabolic functions. The least capable of these functions sets the limit for an individual's physical performance. Of course, other specific capabilities may also cap the possible output: muscular strength is a good example. Furthermore, changes in health, fitness, skill, or motivation affect performance abilities.

Table 8-8 lists the main changes in physiological functions, from rest to maximal work, for ventilation, blood flow, heart rate, and oxygen consumption. Any and all of these variables can be used to assess a person's response to a given workload or to judge the demands imposed by the task. However, only *dynamic* work can be properly assessed by measures of ventilation or metabolic responses. *Static* efforts, in which muscles are contracted and kept so, hinder or completely cut off the blood supply by compression of the capillary bed. Trying to overcome the resistance, the heart increases its contraction rate and the blood pressure; but, because the blood flow remains impeded, relatively little additional energy is consumed. Therefore, such static effort is fatiguing but cannot be assessed well by techniques relying on respiratory or metabolic functions.

Summary (of Section 3)

Energy requirements depend on the activity level of the body to keep the body alive while no effort is made (basal metabolism), to keep the body functioning at rest, and to maintain the body at work.

At the start of work, heart rate increases sluggishly, with oxygen consumption lagging behind it, to finally achieve a steady state. After cessation of the work, heart rate falls back to the resting level, again more quickly than oxygen uptake.

If work demands are too high for the individual, no steady rate can be achieved, and the work cannot be continued without interruption for rest.

Measurement of oxygen consumption (or heart rate) as an indicator of the energetic requirements of the work should be performed during the steady-state phase.

Often, the energy requirement of a job can be calculated by breaking it into elemental subtasks for which the (average) energy requirements are known. In this case, we simply add the energy needs for each subtask.

In some cases, energy requirements can be taken from existing listings. However, the given conditions—for example, those that pertain to work details, climate, or the individual worker—may not be reliably the same as the values that are assumed for the table.

Table 8-8. Change in physiological functions from rest to maximal effort.

Ventilation:	Minute volume, $5 \rightarrow 100$ L/min	×20
	(= Tidal volume × breathing rate)	
	Respiratory frequency, $10 \rightarrow 50$ breaths/min	×5
Blood flow:	Stroke volume, $50 \rightarrow 150$ mL	×3
	Cardiac output: minute volume =	
	stroke vol × HR, $5 \rightarrow 35$ L/min	×7
	Blood pressure	×2
Heart rate:	$60 \rightarrow 180$ beats/min	×3
O_2 consumption:	1 kcal $\rightarrow$ 12 kcal/min	×12

General classifications of job "heaviness" can be established according to ranges of energy and heart rate requirements.

CHAPTER SUMMARY

Calorimetric information can be used for three different purposes:

- To assess individual energetic functions in response to standardized activities: this is a measurement of individual metabolic work capacity.
- To assess the energetic requirements of a given activity as reflected by the energetic responses of a given (standard) subject to that work.
- Combining these two approaches, the energetic requirements of activities can be assessed by measuring the energetic responses of a representative sample of workers.

This information can be used to assess whether task demands are below, at, or above human abilities. If the demands exceed human capabilities, the ergonomist (engineer, manager) should redesign the job by changing tasks, tools, and environment as appropriate and needed; training of the operator in procedures and fitness may also be advisable. This approach to matching requirements with abilities is shown in Figure 8-9.

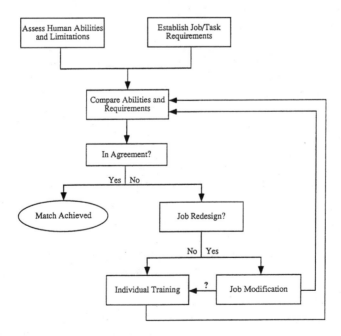

Figure 8-9. Matching task demands with human abilities.

REFERENCES

Astrand, P.O., and Rodahl, K. (1977, 1986). *Textbook of Work Physiology,* 2nd and 3rd ed. New York, NY: McGraw-Hill.

Bernard, T.E., and Joseph, B.S. (1994). Estimation of Metabolic Rate Using Qualitative Job Descriptors. *American Industrial Hygiene Association Journal* 55:11, 1021–1029.

Borg, G.A.V. (1962). *Physical Performance and Perceived Exertion.* Lund, Sweden: Gleerups.

Borg, G.A.V. (1982). Psychophysical Bases of Perceived Exertion. *Medicine and Science in Sports and Exercise* 14:5:377–381.

Durnin, J.V.G.A., and Passmore, R. (1967). *Energy, Work, and Leisure.* London, UK: Heinemann.

Eastman Kodak Company (Vol. 1, 1983; Vol. 2, 1986). *Ergonomic Design for People at Work.* New York, NY: Van Nostrand Reinhold.

Ekman, G. (1964). Is the Power Law a Special Case of Fechner's Law? *Perception and Motor Skills* 19:730.

Erb, B.D., Fletcher, G.F., and Sheffield, T.L. (1979). Standards for Cardiovascular Exercise Treatment Programs. *Circulation,* 59(108A):30–40.

Fechner, G.T. (1860). *Sachen der Psychophysik.* Leipzig, Germany: Breitkopf und Hertel.

Garrow, J.S. (1980). Problems in Measuring Human Energy Balance. In J.M. Kinney (ed.). *Assessment of Energy Metabolism in Health and Disease.* Columbus, OH: Ross Laboratories, pp. 2–5.

Guyton, A.C. (1979). *Physiology of the Human Body,* 5th ed. Philadelphia, PA: Saunders.

Horwat, F., Meyer, J.P., and Malchaire, J. (1988). Validation of a New Pocket Computer Assisted Method for Metabolic Rate Estimation in Field Studies. *Ergonomics* 32:8, 1155–1164.

Kahn, J.F., and Monod, H. (1989). Fatigue Induced by Static Work. *Ergonomics* 32:7, 839–846.

Kinney, J.M. (ed.). (1980). *Assessment of Energy Metabolism in Health and Disease.* Columbus, OH: Ross Laboratories.

Kroemer, K.H.E., Kroemer, H.B., and Kroemer-Elbert, K.E. (1994). *Ergonomics. How to Design for Ease and Efficiency.* Englewood Cliffs, NJ: Prentice-Hall.

Louhevaara, V., Ilmarinen, J., and Oja, P. (1985). Comparison of Three Field Methods for Measuring Oxygen Consumption. *Ergonomics* 28:2:463–470.

Mellerowicz, H., and Smodlaka, V.N. (1981). *Ergometry.* Baltimore, MD: Urban & Schwarzenberg.

Pandolph, K.P. (1983). Advances in the Study and Application of Perceived Exertion. *Exercise and Sport Sciences Review* 11:118–158.

Pennington, A.J., and Church, H.N. (1985). *Food Values of Portions Commonly Used.* New York, NY: Harper.

Roebuck, J.A., Kroemer, K.H.E., and Thomson, W.G. (1975). *Engineering Anthropometry Methods.* New York, NY: Wiley.

Rohmert, W., and Rutenfranz, J. (eds.). (1983). *Praktische Arbeitsphysiologie,* 3rd ed. Stuttgart, Germany: Thieme.

Spitzer, H., and Hettinger, T. (1958). *Tafeln für den Kalorienumsatz bei körperlicher Arbeit.* Darmstadt, Germany: REFA.
Stegemann, J. (1984). *Leistungsphysiologie,* 3rd ed. Stuttgard, Germany: Thieme.
Stevens, S.S. (1957). On the Psychophysical Law. *Psychology Review* 64:151–181.
Webb, P. (1985). *Human Calorimeters.* New York, NY: Praeger.
Weber, E.H. (1834). *De pulse, resorptione, auditu et tactu.* Leipzig, Germany: Kochler.

FURTHER READING

Astrand, P.O., and Rodahl, K. (1986). *Textbook of Work Physiology,* 3rd ed. New York, NY: McGraw-Hill.
Eastman Kodak Company (Vol. 1, 1983; Vol. 2, 1986). *Ergonomic Design for People at Work.* New York, NY: Van Nostrand Reinhold.
Kroemer, K.H.E., Kroemer, H.B., and Kroemer-Elbert, K.E. (1994). *Ergonomics. How to Design for Ease and Efficiency.* Englewood Cliffs, NJ: Prentice-Hall.
Pandolph, K.P. (1983). Advances in the Study and Application of Perceived Exertion *Exercise and Sport Sciences Review.* 11:118–158.
Schmidt, R.F. and Thews, G. (eds.) (1989). *Human Physiology* (2nd ed.). Berlin, Germany; New York, NY: Springer.
Spence, P.A. (1989). *Biology of Human Aging.* Englewood Cliffs, NJ: Prentice-Hall.
Webb, P. (1985). *Human Calorimeters.* New York, NY: Praeger.

Techniques of Indirect Calorimetry

$\mathbf{M}$easuring the oxygen consumed over a sufficiently long period of time is a practical way to assess the metabolic processes. (A physician or physiologist should perform this test.) As discussed earlier, 1 L of oxygen consumed releases about 5 kcal of energy in the metabolic processes. This assumes a normal diet, a healthy body oxidizing primarily carbohydrates and fats under conditions of light to moderate work, and suitable climatic conditions. (The "normality" of the metabolic conditions can be judged, to some degree, by the respiratory exchange quotient, RQ, mentioned earlier.)

Classically, indirect calorimetry has been performed by collecting all exhaled air during the observation period in airtight bags (known as Douglas bags). The volume of the exhaled air is then measured and analyzed for oxygen and carbon dioxide as needed for the determination of the RQ. From these data, the amount of energy used during the collection period can be calculated. This requires a rather complex air collecting system, including nose clip and intake and exhaust valves, which mostly limits this procedure to the laboratory. A major improvement was to divert only a known percentage of the exhaled air into a small collection bag. This procedure (developed in the 1950s in the Max-Planck Institute for Work Physiology) is still in use since only a relatively small device has to be carried by the subject, allowing performance of most daily activities without much hindrance by the collecting system. Still, in both cases, the subject must wear a face mask to measure the total amount of air ventilated and to separate the exhaled from the inhaled or ambient air. This can become quite uncomfortable for the subject and hinders speaking.

Significant advances have been made in the last decades through the use of instantaneously reacting oxygen sensors, which can be placed into the airflow of the exhaled air, allowing a breath-by-breath analysis (Webb, 1985). The differences in oxygen measured with different equipment are usually rather small; for example, a comparison between the classical Douglas bag method, the Max-Planck gas meter, and the Oxylog®, a small portable instrument, showed variations in the mean of less than 7%; the linear regression coefficients were better than 0.90 (Louhevaara, Ilmarinen, and Oja, 1985; Horwat, Meyer, and Malchaire, 1988). For most field observations, the accuracy of the bagless procedures is quite sufficient. Since the volume of exhausted air can also be measured by suitable sensors, more recently, "open"-face masks have been developed that draw a stream of air across the face of the subject, who simply inhales from this airflow and exhales into it. This allows free breathing and speech and even cools the face (which might influence the working capacity to a small extent).

Rating the Perceived Effort

Early in the 19th century, models of the relationships between a physical stimulus and one's perceptual sensation of that stimulus—that is, the "psychophysical correlate,"—were developed. Weber (1834) suggested that the "just noticeable difference ΔI" that can be perceived depends on the absolute magnitude of the physical stimulus I:

$$\Delta I = a\,I \qquad (8\text{-}7)$$

where a is constant.

Fechner (1860) related the magnitude of the "perceived sensation P" to the magnitude of the stimulus I:

$$P = b + c \log I \qquad (8\text{-}8)$$

where b and c are constants.

In the 1950s, S.S. Stevens at Harvard and G. Ekman in Sweden introduced ratio scales (assuming a zero point and equidistant scale values) that have since been used to describe the relationships between the perceived intensity I and the physically measured intensity P of a stimulus in a variety of sensory modalities (e.g., related to sound, lighting, and climate):

$$P = d\,I^{n} \qquad (8\text{-}9)$$

where d is a constant and n ranges from 0.5 to 4, depending on the modality.

Since about 1960, Borg and his collaborators have modified these relationships to take into account deviations from previous assumptions (such as zero point and equidistance) and to describe the perception of different kinds of physical efforts. Borg's *general function* is

$$P = e + f(I - g)^n \tag{8-10}$$

with the constant e representing "the basic conceptual noise" (normally below 10% of I) f is a conversion factor dependent on the type of effort and the constant g indicate the starting point of the curve.

Ratio scales indicate only proportions between percepts but do not indicate absolute intensity levels. They neither allow intermodal comparisons nor comparisons between intensities perceived by different individuals. Borg has tried to overcome this problem by assuming that the subjective range and intensity level are about the same for each subject at the level of maximum intensity. In 1960, this led to the development of a category scale for the rating of perceived exertion (RPE). The scale ranges from 6 to 20 (to match heart rates from 60 to 200 bpm). Every second number is anchored by verbal expressions:

The 1960 Borg RPE Scale (Modified 1985):
6—(no exertion at all)
7—extremely light
8
9—very light
10
11—light
12
13—somewhat hard
14
15—hard (heavy)
16
17—very hard
18
19—extremely hard
20—maximal exertion

In 1980, Borg proposed a category scale with ratio properties, which could yield ratios and levels and allow comparisons but still retain the same correlation (about 0.88) with heart rate as the RPE scale, particularly if large muscles were involved in the effort.

The Borg General Scale (1980):
0—nothing at all
0.5—extremely weak (just noticeable)
1—very weak
2—weak

3—moderate
4—somewhat strong
5—strong
6
7—very strong
8
9
10—extremely strong (almost maximal)

(Note: The terms *weak* and *strong* may be replaced by *light,* and *hard,* or *heavy,* respectively).

The instructions for use of the scale are as follows (modified from Borg's publications). While the subject looks at the rating scale, the experimenter says:

"I will not ask you to specify the feeling, but do select a number that most accurately corresponds to your perception of (experimenter specifies symptom).

If you don't feel anything, for example, if there is no (symptom), you answer *zero*—nothing at all.

If you start feeling something just barely noticeable, you answer 0.5—extremely weak, just noticeable.

If you have an extremely strong feeling of (symptom), you answer 10—extremely strong, almost maximal. This is the absolute strongest that you have ever experienced.

The more you feel, the stronger the feeling, the higher the number you choose.

Keep in mind that there are no wrong numbers. Be honest. Do not overestimate or underestimate your ratings. Do not think of any other sensation than the one I ask you about.

Do you have any questions?"

Let the subject get well acquainted with the rating scale before the test. During the test, let the subject do the ratings toward the end of every work period, that is, about 30 sec before stopping or changing the workload. If the test must be stopped before the scheduled end of the work period, let the subject rate the feeling at the moment of stoppage.

Chapter 9

Interactions of the Body with the Thermal Environment

OVERVIEW

This chapter discusses a variety of thermal interactions between the body and its environment. This includes heat generation within the body in relation to the energy emitted from it via external work done and by evaporation and the heat exchanged (received or dissipated) by radiation, convection, and conduction. The physiological and physical means to affect this energy transfer and the engineering control of the macro- and microclimate are explained. Recommendations for suitable ergonomic conditions are given.

THE MODEL

The human body generates energy and exchanges (gains or loses) energy with the environment. Since a rather constant core temperature must be maintained, suitable heat flow from the environment to the body, or from the body to the environment, must be achieved. The internal energy flow is controlled primarily in the body masses between skin and core.

INTRODUCTION

Two overriding facts explain the main aspects of the heat flow within the body and between the body and the environment:

- Body core temperature must be maintained close to 37°C.
- Heat flows from the warmer to the colder matter.

THE HUMAN BODY IS A THERMO-REGULATED SYSTEM

The human body has a complex control system to maintain the body core temperature near 37°C (about 99°F) with a variation of just a few degrees. There is some fluctuation throughout the day and from day to day due to diurnal and other changes in body functions. However, the main impact on the human thermal regulatory system results from the interaction between (metabolic) heat generated within the body and external energy gained in hot surroundings or lost in a cool environment. If the deep body temperature deviates just a few degrees from its set value, physical and mental work capacities are impaired: cellular structures, enzyme systems, and many other functions are directly affected by changes in body temperature.

If the temperature at a human cell exceeds 45°C, heat coagulation of proteins takes place. If the temperature reaches freezing, ice crystals break the cell apart. In its effort to protect itself from conditions that are either too hot or too cold, the body uses a temperature regulation system to keep temperatures well above freezing and below the 40s in its outer layers. At the core, a range close to 37°C must be maintained; changes in core temperature of ± 2°C from 37°C affect body functions and task performance severely, while deviations of ± 6°C are usually lethal (ASHRAE, 1992a, b).

The Energy Balance

In Chapter 8 which dealt with the metabolic system, the energy equation (8-1) between body energy inputs and outputs was given as

$$I = M = H + W + S^{1}$$
(9-1)

where
  I = the energy input via nutrition, which results in the body's rate of free
      metabolic energy production M
  W = the rate of external work done
  H = the rate of heat that must be released from the body
  S = the rate of energy storage in the body

All rates are in watts per square meter of body surface, which is nearly 2 m^2 in men.

Assuming, for convenience, that the quantities I and W remain unchanged, we can concentrate on the energy exchange with the thermal environment:

$$I - W = \text{const} = H + S$$
(9-2)

---

[1]This is the terminology used by Parsons (1993) which is different from the use in the previous two editions of this book.

The system is in balance with the environment if all heat energy H is dissipated to the environment without a change in the quantity S. If heat storage S increases, not all metabolic energy could be dispelled to the environs and/or energy was transferred from the environment to the body. If S becomes smaller, more than H must have been lost from the body to the environment.

ENERGY EXCHANGES WITH THE ENVIRONMENT

Energy is exchanged with the environment through radiation, convection, conduction, and evaporation.

Heat exchange through *radiation R* depends primarily on the temperature difference between two opposing surfaces, for example, between a windowpane and a person's skin. Heat is always radiated from the warmer to the colder surface; for example, to the cold window in the winter or to the body from a sun-heated pane in the summer. Therefore, the body can either lose or gain heat through radiation. This radiative heat exchange does not depend on the temperature of the air between the two opposing surfaces.

The amount of radiating energy Q_R (in J/s) gained $(+)$ or lost $(-)$ by the human body through radiation is

$$Q_R = a S (dT_O^4 - eT^4) \qquad (9\text{-}3)$$

where

a = radiation constant in 75 J/(sm^2°K^4)
S = body surface participating in the energy exchange, in m^2
d = absorption coefficient (see below)
T_O = temperature of opposing surface, in degrees K
e = emission coefficient (see below)
T = body surface temperature, in degrees K.

Equation (9-3) is a form of the Stephan–Bolzmann law of radiative heat transfer.

The wavelengths of radiation from the human body are $3 < \lambda < 60$ μm, that is, in the infrared range. Hence, it radiates like a black body, with an emission coefficient e close to 1, independent of the color of the radiating human skin. However, the absorption coefficient d depends on skin color, for solar rays (with wavelengths $0.3 < \lambda < 4$ μm) it ranges from 0.6 for light-skinned people to 0.8 for dark-skinned people.

Energy is also exchanged through *convection C* and *conduction K*. In both cases, the heat transferred is proportional to the area of human skin participating in the process and to the temperature difference between skin and the adjacent layer of the external medium. In general terms, heat exchange per second (by convection or conductance) is

$$Q_{C,K} = f[Sh(t_m - t)] \qquad (9\text{-}4)$$

where
h = heat conduction coefficient (see below)
S = body surface participating in the heat exchange, in m^2
t_m = temperature of the medium with which S is in contact, in °C
t = temperature of the body surface S, in °C.

Equation (9-4) is a form of Newton's law of cooling.
The heat conduction coefficient h of human tissue is $3 < h < 260$ J/(cms°C), that is, the amount of energy that penetrates 1-cm-thick tissue per sec when the temperature difference is 1°C.

Conductance K exists when the skin contacts a solid body, such as a piece of iron. Energy flows from the warmer body to the colder one; as the temperatures of the contact surface become equal, the energy exchange ceases. The rate and amount of heat exchanged also depend on the conductance of the touching bodies. Cork or wood feel warm because their heat conduction coefficients are below that of human tissue, but cool metal accepts body heat easily and conducts it away.

Exchange of heat through *convection C* takes place when the human skin is in contact with air and fluids, such as water. As in conduction, heat energy from the skin is transferred to a colder gas or fluid next to the skin surface or transferred to the skin if the surrounding medium is warmer. Convective heat exchange is facilitated if the medium moves quickly along the skin surface (in laminar or, more often, turbulent fashion), thus maintaining a temperature differential.

The method of exchanging heat via convection Q_C (gain indicated by + and loss by –) is similar to conduction; but the effect of moving the immediate layer of the surrounding medium modifies the process:

$$Q_C = c\, S\, (t_m - t) \tag{9-5}$$

with the convection coefficient, in general, $21 < c < 37$ kJ/(hrm^2°C) (Stegemann, 1984). It depends heavily on the actual relative movement of the medium. As long as there is a temperature gradient between the skin and the medium, there is always some natural movement of air or fluid: this is called *free convection*. Much more movement can be produced by forced action (e.g., by an air fan or while a person is swimming in water rather than floating motionless): this is called *induced convection*.

For the nude body in water, the heat loss is (Nadel, 1984)

$$Q_C = h_c\, S\, (t_m - t) \qquad t_m < t \tag{9-6}$$

with the convective transfer coefficient h_c at about 230 W/(m^2 °C) during rest in still water; it is about 580 W/(m^2 °C) for a person swimming at any speed because of the turbulence of the water layer near the body produced by the swimming activities.

Heat exchange by *evaporation E* is in only one direction. Human beings lose heat by evaporation; there is no condensation of water on the skin, which would add heat. Evaporation of water (sweat) on the skin requires an energy of about 2440 J/cm³ (580 cal/cm³) of evaporated fluid, which reduces the heat content of the body by that amount.

The heat lost by evaporation Q_E from the human body is a function of participating wet body surface, humidity, and vapor pressures (according to Nadel and Horvath, 1975)

$$Q_E = f[S(h_r p_a - p)] \qquad (9\text{-}7)$$

where
S = body surface participating in the heat dispersion
h_r = relative humidity of the surrounding air
p_a = vapor pressure in the surrounding air
p = vapor pressure at the skin

The units for pressure are 1000 Nm⁻² = 1 kPa = 10 mb = 7.52 mm Hg = 7.52 Torr.

Of course, Q_E is zero for $p < p_a$ since there can only be heat loss by evaporation if the surrounding air is less humid than the air directly at the skin. Therefore, movement of the air layer at the skin (convection) increases the actual heat loss through evaporation if this replaces humid air by drier air.

Some evaporative heat loss occurs even in a cold environment because there is always evaporation of water in the lungs that increases with enlarged ventilation at work, and there is also secretion of some sweat onto the skin surface in physical work. The nude body at rest in the cold loses 6 to 10 W/m² from the skin and 3 to 6 W/m² from the respiratory tract (Nadel and Horvath, 1975).

Given these variables, heat balance exists when heat H developed in the body, heat storage S in the body, and heat exchanges with the environment by radiation R, convection C, conduction K, and evaporation E are in equilibrium. This can be expressed as

$$H + S + R + C + K + E = 0 \qquad (9\text{-}8)$$

The quantities R, C, K, and E are counted as negative if the body loses energy to the environment and positive if the body gains energy from the environment. Recall that E can only be negative.

TEMPERATURE REGULATION AND SENSATION

Heat energy is circulated throughout the body by the blood. The blood flow is modulated by the vasomotor actions of constriction, dilation, and shunting. Heat is exchanged with the environment at the body's respiratory surfaces, that is, in the lungs and at respiratory mucosa and, of course, through the skin.

Heat is produced in the body's metabolically active tissues: primarily at skeletal muscles but also in internal organs, fat, bone, and connective and nerve tissue. *In a cold environment, heat must be conserved,* which is accomplished primarily by the reduction of blood flow to the skin and by increased insulation. *In a hot environment, body heat must be dissipated, and gain from the environment must be prevented,* primarily by increased blood flow to the skin, by sweat production, and by evaporation.

The body must regulate its temperature to prevent undercooling or overheating. The temperature of key tissues, such as brain, heart, lungs, and abdominal organs, must be kept rather constant. However, the body temperature is not at all uniform; there are large temperature differences between the "core" and the "shell." Under normal conditions, the average gradient between skin and deep body is about 4°C at rest but, in the cold, the difference in temperature may be 20°C or more. Even within the core, the temperature varies by several degrees. This makes it rather difficult to speak about "one" body temperature since it is so variable throughout the body. In reality, the temperature regulation system has to maintain various temperatures at various locations under different conditions.

Figure 9-1 shows a model of the regulation of the human energy balance. It breaks the actual regulatory system into three subsystems: the controlling, the effecting, and the regulated elements. The human body has given set points, near 37°C in the brain and about 33°C at the skin. Any deviations from these values

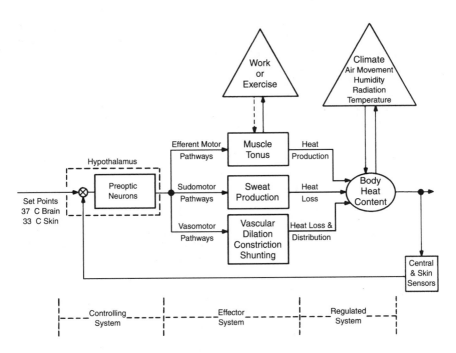

Figure 9-1. Model of the regulation of body heat content.

are detected by various temperature sensors, and counteractions are initiated at the hypothalamus. Here, neurons affect three different pathways: the efferent nervous system to change the muscle tonus, the sudomotor system to affect sweat production, and the vasomotor system to bring about vascular dilation, constriction, or shunting. The effecting system interacts with the work or exercise being performed. For example, if less heat must be generated internally, muscular activities will be reduced, possibly to the extent that no work is being performed anymore. On the other hand, if more heat must be generated the work or exercise level will be augmented by increased muscular activities. (Given the low efficiency of muscular work, it generates much heat.) However, muscle activities can generate only more or less heat but cannot cool the body. In contrast, sweat production influences only the amount of energy lost but cannot bring about a heat gain. Vascular activities can affect the heat distribution through the body and control heat loss or gain but do not generate energy.

Muscular, vascular, and sweat production functions regulate the body heat content in direct interaction with the external climate. The climate itself is defined by humidity, radiation, temperature, and air (or fluid) movement, as discussed later.

Various temperature sensors are located in the core and the shell of the body. Hot sensors generate signals (sent to the hypothalamus), particularly in the range of approximately 38° to 43°C. On the other hand, the major sensitivity to cold ranges from about 15° to 35°C. There is some overlap in the sensations of "cool" and "warm" in the intermediate range. Between about 15° and 45°C, our perception of either "cold" or "hot" condition is highly adaptable. Below 15°C and above 45°C, the human temperature sensors are less discriminating but also less adapting. A paradoxical effect is that, around 45°C, sensors again signal "cold" while, in fact, the temperature is rather hot.

Deviations in the actual temperatures from the set points (in the brain and skin) bring about counteractions by the controlling system. However, the set points are variable by several degrees and change, for example, throughout the course of the day with diurnal rhythms or with acclimatization.

THERMAL HOMEOSTASIS

The human regulatory system must achieve two suitable temperature gradients: from the core to the skin and from the skin to the surroundings. The gradient from the core to the skin is the more important of the two because overheating or undercooling of the key tissues in the brain and the trunk must be avoided, even at the cost of overheating or undercooling the shell. In the central body parts, this gradient should not exceed approximately 4°C.

The core/skin gradient is achieved in the more peripheral tissues, mostly in the network of superficial arteries and veins and their connections as means to transport and distribute heat. Thermal homeostasis is achieved *primarily* by regulation of the blood flow from deep tissues and muscles to lungs and skin.

The ability to absorb and transport heat is 3 to 3.5 J/(g°C) in tissue but reaches 4 J/(g°C) in the blood. This shows that blood is a relatively efficient way to transport heat within the body. Of the total dissipated heat, most is transmitted to the environment through the skin; the remainder is exchanged through the lungs.

Secondary activities to establish thermal homeostasis take place at the muscles either by involuntary shivering or by voluntary changes in activities, with concurrent changes in internal heat generation, given the energetic inefficiency of the body. Different actions are taken depending on the goal of the regulatory system. If more heat is needed, skeletal muscle contractions are initiated. If too much heat is generated in the body, muscular activities are abolished.

Changes in clothing and shelter are *tertiary* actions to achieve thermal homeostasis. They affect radiation, convection, conduction, and evaporation. "Light" or "heavy" clothes have different permeability and ability to establish stationary insulating layers. Clothes affect conductance, that is, energy transmitted per surface unit, time, and temperature gradient. Also, their color determines how much external radiation energy is absorbed or reflected. Similar effects are brought about by shelters which, by their material, distance from the body, form and color, determine whether heat is gained or lost by the body through radiation, convection, and evaporation.

MEASUREMENT OF BODY TEMPERATURES

One way to assess the exchange of heat with the environment is to perform direct calorimetry, where a person is placed into an energy-tight compartment which allows the measurement of all heat energies exchanged; see Section 2 in Chapter 8. This is a tedious procedure, however, which severely limits the ability of the person to work "normally." Most methods to assess the heat balance are based on temperature measurements within or at the body.

The so-called safe temperature range of the core is between 35° to 40°C. Impairment of functions in the core, especially of brain and heart occur below 33 and above 42°C in these organs. Changes in core temperatures, particularly severe deviations from the "safe" temperatures, indicate work and environmental overloads or dysfunctions of the energy regulatory mechanisms in the body.

A variety of techniques exist to measure the temperature in body parts. A "classical" site is the rectum, where temperature probes are usually inserted 5 to 10 cm behind the sphincter. With the hypothalamus providing the reference temperature, the rectal temperature is usually about 0.5°C lower, provided that a steady state has been maintained for about 30 min. The rectal temperature of a resting individual is slightly higher than the temperature of arterial blood and about the same as in the liver. Brain temperature rises more quickly in response to heat influx than rectal temperature. Temperatures measured at the eardrum follow actual brain temperatures rather closely, as do the temperatures measured in the esophagus. Inserting a temperature probe into the esophagus or the stomach allows measurement of deep body temperatures. Temperature measurement in the mouth or in the armpit is less accurate but rather easily done and is more acceptable.

As discussed earlier, skin temperatures may vary much from core temperatures; they can be 20°C below or 10°C above the temperatures at the hypothalamus. Measurement of temperatures at various skin sites, therefore, provides information that is only loosely related to core temperatures. Furthermore, various sites on the body surface may be at very different temperatures. Hence, the concept of "average skin temperature" is a difficult one; however, it has been used by assigning weighting factors to the measurements taken at various body surfaces, depending on the proportions of these surface areas compared to the total body. For example, the Hardy and Dubois procedure is to multiply measurements taken at the head by 0.07, the arms by 0.14, the hands by 0.05, the trunk by 0.35, the thighs by 0.19, the lower legs by 0.13, and at the feet by 0.07. The results are added for an "average" skin temperature.

Various procedures exist to measure and calculate the mechanisms of heat regulation. For example, if we establish the "average" skin temperature T_s, measure the rectal temperature T_r, and know the body mass W in kilograms, we can calculate the heat content from the Burton equation,

$$\text{Heat content} = 3.47 \, W \, (0.65 \, T_r + 0.35 \, T_s). \tag{9-9}$$

However, the specific heat of the body, for this equation assumed to be 3.47 kJ/(kg°C), may vary from 2.93 to 3.45, depending on the individual's body composition. Also, the ratios of T_r and T_s, assumed to be 0.65 and 0.35, respectively, are not fixed but range from 0.9 and 0.1 in warm environments and during exercise to 0.6 and 0.4 at rest in a cool environment (Stolwijk, 1980).

ASSESSMENT OF THE THERMAL ENVIRONMENT

The thermal environment is determined by four physical factors: air (or water) temperature, humidity, air (or water) movement, and temperature of body surfaces (Malchaire, 1995; Olesen and Madsen, 1995). The combination of these four factors determines the physical conditions of the climate and our perception of the climate.

Measurement of *temperature* is commonly performed with thermometers filled with alcohol or mercury; thermistors or thermocouples can be used as well. Whichever technique is used, it must be ensured that the ambient air temperature is not affected by the other three climate factors (humidity, air movement, and surface temperatures). To measure the so-called dry temperature of air, one keeps the sensor dry and shields it with a surrounding bulb that reflects radiated energy. Hence, air temperature is properly measured with a so-called dry-bulb thermometer and often called *dry temperature.*

Air humidity may be measured with a hygrometer: originally, a human or horse hair that changed its length with wetness, now an instrument whose electrical conductivity (resistance/capacitance) alters with the existing humidity. Another instrument is the psychrometer, which uses one dry and one wet thermometer. The dry thermometer usually shows a higher temperature than the wet thermometer, which is cooled by evaporation. The difference between the temperatures increases with more evaporation, i.e., with less humidity. A psychrometer is called *natural* if there is no artificial air movement and called *forced* if there is.

Air humidity may be expressed either in absolute or in relative terms. The highest absolute content of vapor in the air is reached when any further increase would lead to the development of water droplets falling out of the gas. This dew point depends on air temperature and barometric pressure: higher temperature and pressure allow more water vapor to be retained than lower conditions. One usually speaks of relative humidity, which indicates the actual vapor content in relation to the possible maximal content at the given air temperature and air pressure.

Air movement is measured with various types of anemometers, usually based on mechanical or electrical principles. One may also measure air movement with two thermometers, one dry and one wet (similar to what can be done to assess humidity), relying on the fact that the wet thermometer shows more increased evaporative cooling with higher air movement than the dry thermometer. Air movement helps particularly in convective heat exchange because it moves "fresh" air to skin surfaces. Here, turbulent air movement is as effective as laminar movement in heat transfer.

Radiant heat exchange depends primarily on the difference in temperatures between the individual and the surroundings, on the emission properties of the radiating surface, and on the absorption characteristics of the receiving surface. While there is no problem in measuring *surface temperatures,* one easy way to assess the amount of energy transferred through radiation is to place the thermometer inside a black globe, which absorbs practically all radiated energy.

Various techniques exist to express the combined effects of the four environmental factors in one model, chart, or index (Eissing, 1995; Mairiaux and Malchaire, 1995). For example, the outdoors WBGT (wet-bulb globe temperature) weighs the effects of several climate parameters:

$$\text{WBGT (outdoors)} = 0.7\,\text{WB} + 0.2\,\text{GT} + 0.1\,\text{DB} \qquad (9\text{-}10)$$

where WB is the wet-bulb temperature of a sensor in a wet wick exposed to natural air current; GT is the globe temperature at the center of a black sphere of 15-cm diameter; and DB is the dry-bulb temperature measured while shielded from radiation.

The indoors WBGT is simpler:

$$\text{WBGT (indoors)} = 0.8\,\text{WB} + 0.3\,\text{GT} \qquad (9\text{-}11)$$

The WBGT is commonly applied to assess the effects of warm or hot climates (Parsons, 1995). Depending on the activity level, expressed in watts, the WGBT temperatures given in Table 9-1 are considered "safe" for most healthy people although there is some concern about the adequacy of the WGBT for combinations of high humidity with little air movement (Ramsey, 1990, 1995)

Various instruments on the market automatically combine several measurements into one index number, including the effective temperature, discussed later.

REACTIONS OF THE BODY TO COLD ENVIRONMENTS

The human body has few natural defenses against a cold environment. Most of the actions taken are behavioral in nature, such as putting on suitably heavy clothing

Table 9-1. "Safe" WGBT values (°C).

M, metabolic rate*	Person is heat-acclimatized	Person is not acclimatized
$M \leq 117$ W	33	32
$117 < M \leq 234$ W	30	29
$234 < M \leq 360$ W	28	26
$360 < M \leq 468$ W	No air movement: 25 With air movement: 26	No air movement: 22 With air movement: 23
$M > 468$ W	No air movement: 23 With air movement: 25	No air movement: 18 With air movement: 20

*Assuming a skin surface area of 1.8 m².

Source: Abbreviated from ISO 7243, 1982.

covering the skin (Lotens et al., 1995), seeking shelter, or using external sources of warmth.

In a cold climate, the body must conserve heat while producing it. For this, there are two major ways to regulate the temperature: redistribution of the blood flow and increase in metabolic rates.

Redistribution of Blood

To conserve heat, the temperature of the skin is lowered to reduce the temperature difference against the outside. This is done by displacing the circulating blood toward the core, away from the skin. This can be rather dramatic; for example, the blood flow in the fingers may be reduced to 1% of what existed in a moderate climate.

Blood distribution can be regulated by three procedures: constriction of skin vessels, use of deep veins, and increased heat exchange between arteries and veins.

In a lightly clad, resting individual, in an "ideal" external temperature of about 28°C, the mean skin temperature is about 33°C and the core temperature about 37°C. This temperature gradient, from core to skin, allows the transfer of excess heat from the metabolically active tissues to the environment. Of the total circulating blood volume (about 5 L), about 5% flows through the blood vessels in the skin. A cold environment constitutes a large temperature difference to the skin, which would cause increased heat loss through convection and radiation. By closing much of the blood vessel pathways, less blood flows toward the superficial skin surfaces; the skin temperature is lowered and, hence, the energy flow toward the environment is reduced. One may consider this a reduction in the conductance of the surface tissues. Activation of cutaneous vasoconstriction is apparently under the control of the sympathetic nervous system (see Chap. 4), in addition to local reflex reactions to direct cold stimuli.

An interesting phenomenon associated with cutaneous vasoconstriction is the "hunting reflex," a cold-induced vasodilation: after initial vasoconstriction has taken place, a sudden dilation of blood vessels occurs that allows warm blood to re-turn to the skin—of the hands, for instance—which rewarms that section of the body. Then, vasoconstriction returns, and this sequence may be repeated several times. If vasoconstriction and metabolic rate regulation cannot prevent serious en-ergy loss through the body surfaces, the body will suffer some effects of cold stress.

Vasoconstriction of the skin blood vessels is usually accompanied by the sec-ond method for reducing tissue conductance: blood in the veins of the extremities and near the skin is rerouted from the superficial to the deep veins. These deep veins are anatomically close to the arteries, which carry warm blood from the heart. Therefore, a heat exchange between the arteries and veins in the deep body tissues occurs. Having cooled arterial blood supplying the skin or extremities brings about two effects: cooling of the body core is reduced, and the extremities and surfaces are cool, with decreased heat conductance to the outside.

This displacement of the blood volume from the skin to the central circula-tion is very efficient in keeping the core warm and the surfaces cold. Peripheral vasoconstriction can bring about a sixfold increase in the insulating capacity of the subcutaneous tissues, accompanied by the earlier-mentioned reduction of the blood volume to as little as 1% of normal. The danger associated with these vaso-constrictional regulatory actions is that the temperature in the peripheral tissues may approach that of the environment. Thus, cold fingers and toes may result, with possible damage to the tissue if the temperatures get close to freezing. The blood vessels of the head do not undergo as much vasoconstriction, and so the head stays warm even in cold environments, with less danger to the tissues; the resulting large difference in temperature to the environment, however, brings about a large heat loss, which can be prevented by wearing a hat or scarf to create an insulating layer.

The mechanism of blood redistribution in the cold indicates, again, the over-riding need to keep the core temperature high enough, even at the risk of cooling the shell to the extent that local damage may occur there.

Increased Metabolic Heat Production

The other major reaction of the body to a cold environment is the increase in metabolic heat generation. This may occur involuntarily, by shivering (*thermo-genesis*). Shivering usually begins in the neck, apparently to warm the critically important flow of blood to the brain. Its onset is normally preceded by an increase in overall muscle tone in response to body cooling. With increased firing rates of motor units (see Chap. 3) but no actual movements generated, a feeling of stiff-ness is generally experienced. Then, suddenly, shivering begins, caused by mus-cle units firing at different frequencies of repetition (rate coding; see Chap. 3) and out of phase with each other (recruitment coding). Since no mechanical work is done to the outside, the total activity is transformed into heat production, allow-ing an increase in the metabolic rate to up to four times the resting rate. If the body does not become warm, shivering may become rather violent when motor unit innervations become synchronized, so that large muscle units are contracted. While such shivering can generate heat that is five or more times the resting metabolic rate, it can be maintained for only a short period at a time. There may be another mechanism to produce heat, called *nonshivering thermogenesis:* body organs, particularly in the liver and the viscera, increase their metabolism. The existence of this response in humans is debated.

Of course, muscular activities can also be voluntary, such as increasing the dynamic muscular work performed or moving body segments, contracting muscles, flexing the fingers, and so forth. Since the energy efficiency of the body is very low (see Chap. 8), dynamic muscular work may easily increase the generation of metabolic heat to 10 or more times the resting rate.

Incidentally, the development of "goose bumps" of the skin helps to retain a layer of stationary air close to the skin, which is relatively warm and has the effect of an insulating envelope, reducing energy loss at the skin.

How Cold Does It Feel?

In a cold environment, an individual's decision to stay in the cold or to seek shelter depends on the subjective assessment of how cold body surfaces or the body core actually is. It is a dangerous situation if a person fails to perceive and to react to the body's signals that it is becoming dangerously cold or if the body temperature becomes so low that further cooling is below the threshold of perception.

The perception of the body getting cold depends on signals received from surface thermal receptors, from sensors in the body core, and from some combination of these signals. As skin temperatures decrease below 35.5°C, the intensity of the cold sensation increases; cold sensation is strongest near 20°C but, at lower temperatures of perception, the intensity decreases. It is often difficult to separate feelings of cold from pain and discomfort.

The conditions of cold exposure may greatly influence the perceived coldness. It can make quite a difference whether one is exposed to cold air (with or without movement) or to cold water, whether or not one wears protective clothing, and what one is actually doing. When the temperature plunges, each downward step can generate an "overshoot" sensation of cold sensor receptors that react very quickly, not only to the difference in temperature but also to the rate of change. Yet, if the temperature stabilizes, the cold sensations become smaller as one adapts to the condition. Exposure to very cold water accentuates the overshoot phenomenon observed in cold air. This may be due to the fact that the thermal conductivity of water is about a thousand times greater than that of cold air at the same temperature. Thus, cold water causes a convective heat loss that may be 25 times that of cold air. In experiments, subjects (wearing a flotation suit) were immersed in cold water of 10°C. Their temperatures at groin, back, and rectum were continuously recorded, and the subjects rated how cold they perceived these areas to be. The results of the experiment showed that the subjects were unable to assess reliably how cold they actually were. Neither their core nor surface temperatures correlated with their cold sensations (Hoffman and Pozos, 1989).

Altogether, the results of many experiments and experiences indicate that the subjective sensation of cold is a poor, possibly dangerous indicator of core and surface temperature of the body. Measuring ambient temperature, humidity, air movement, and exposure time and reacting to these physical measures probably constitute a better strategy than relying on subjective sensations.

Indices of Cold Strain

If vasoconstriction and metabolic rate regulation cannot prevent serious energy loss through the body surfaces, the body will suffer some effects of cold stress. The skin is, as just discussed, first subjected to cold damage, while the body core is protected as long as possible.

As the skin temperature is lowered to about 15° to 20°C, manual dexterity begins to decrease. Tactile sensitivity is severely diminished as the skin temperature falls below 8°C. If the temperature approaches freezing, ice crystals develop in the cells and destroy them, a result known as frostbite. Reduction of core temperature is more serious, where vigilance may begin to drop at temperatures below 36°C. At core temperatures of 35°C, one may be unable to perform even simple activities. When the core temperature drops even lower, the mind becomes confused, with loss of consciousness occurring at around 32°C. At core temperatures of about 26°C, heart failure may occur. At very low core temperatures, such as 20°C, vital signs disappear, but the oxygen supply to the brain may still be sufficient to allow revival of the body from hypothermia.

Severe reductions in skin temperatures are accompanied by a fall in core temperature. At local temperatures of 8° to 10°C, peripheral motor nerve velocity is decreased to near zero; this generates a nervous block, which helps to explain why local cooling is accompanied by rapid onset of physical impairment. Severe cooling of the skin and central body goes along with increasing inability to perform activities, even if they could save the person ("cannot light a match") leading to apathy ("let me sleep") and final hypothermia.

Hypothermia can occur very quickly if a person is exposed to cold water. While one can endure up to 2 hr in water at 15°C, one is helpless in water of 5°C after 20 to 30 min. The survival time in cold water can be increased by wearing clothing that provides insulation; also, obese people with much insulating adipose tissue are at an advantage over skinny people. Floating motionless results in less metabolic energy generated and spent than when swimming vigorously.

REACTIONS OF THE BODY TO HOT ENVIRONMENTS

In hot environments, the body produces heat and must dissipate it by convection, conduction, radiation, and evaporation. As in cold environments, two primary means control the energy flow: blood distribution and metabolic rate. Now, however, the body must dissipate heat instead of preventing heat loss. To achieve this, the skin temperature should be near or, better yet, above that of the immediate environment.

Redistribution of Blood

Blood is redistributed to facilitate heat transfer to the skin. For this, the skin vessels are dilated and the superficial veins fully opened, actions directly contrary to the ones taken in the cold. This may bring about a fourfold increase in blood flow above the resting level, increasing the conductance of the tissue. Accordingly, energy loss through convection, conduction, and radiation (which all follow the temperature differential between skin and environment) is facilitated.

If heat transfer is still not sufficient, sweat glands are activated, and the evaporation of the produced sweat cools the skin. Recruitment of sweat glands from different areas of the body varies among individuals. Some persons have few sweat glands, while most have at least 2 million sweat glands in the skin, so that large differences in the ability to sweat exist among individuals. The activity of each sweat gland is cyclic. The overall amount of sweat developed and evaporated depends very much on clothing, environment, work requirements, and the individual's acclimatization.

Reduction of Muscle Activities

If heat transfer by blood distribution and sweat evaporation is insufficient, muscular activities must be reduced to lower the amount of energy generated through metabolic processes. In fact, this is the final and necessary action of the body if, otherwise, the core temperature would exceed a tolerable limit. If the body has to choose between unacceptable overheating and continuing to perform physical work, the choice will be in favor of core temperature maintenance, which means reduction or cessation of work activities.

Indices of Heat Strain

There are several signs of excessive heat strain on the body. The first one is the sweat rate. Above the so-called insensible perspiration (in the neighborhood of about 50 cm^3/hr) sweat production increases depending on the heat that must be dissipated. In strenuous exercises and hot climates, several liters of sweat may be produced in 1 hr. On the average, however, during the working time, usually not more than about 1 L/hr is produced, but sweat losses up to 12 L in 24 hr have been reported under extreme conditions. Sweat begins to drip off the skin when the sweat generation has reached about one-third of the maximal evaporative capacity. Of course, sweat running down the skin contributes very little to heat transfer.

Increases in the circulatory activities signal heat strain. Cardiac output must be enlarged, which is mostly brought about by a higher heart rate. This may be associated with a reduction in systolic blood pressure. Another sign of heat strain is a rise in core temperature which must be counteracted before the temperature exceeds the sustainable limit.

The water balance within the body provides another sign of heat strain. Dehydration indicated by the loss of only 1% or 2% of body weight can critically affect the ability of the body to control its functions. Therefore, the fluid level must be maintained; the best way is to drink frequently small amounts of water. Sweat contains different salts, particularly NaCl, in smaller concentrations than in the blood. Sweating, which extracts water from the plasma, augments the relative salt content of the blood. Normally, it is not necessary to add salt to drinking water since, in Western diets, the salt in the food is more than sufficient to resupply the salt lost with the sweat.

Water supply to the body comes from fluids drunk, water contained in food, and water chemically liberated during oxidation of nutrients. Approximate daily water losses are: from the gastrointestinal tract, 0.2 L; from the respiratory tract, 0.4 L; through the skin, 0.5 L; from the kidneys, 1.5 L. Obviously, these figures can change considerably when a person performs work in a hot environment.

Among the first reactions to heavy exercise in excessive heat are sensations of discomfort and perhaps skin eruptions ("prickly heat") associated with sweating. As a result of sweating, so-called heat cramps may develop, which are muscle spasms related to local lack of salt. They may also occur after quickly drinking large amounts of fluid.

Heat exhaustion is a combined function of dehydration and overloading the circulatory system. Associated effects are fatigue, headache, nausea, dizziness, often accompanied by giddy behavior. Heat syncope indicates a failure of the circulatory system, demonstrated by fainting. Heat stroke indicates an overloading of both the circulatory and sweating systems and is associated with hot, dry skin; increased core temperature; and mental confusion. Table 9-2 lists symptoms, causes, and treatment of heat stress disorders.

Table 9-2. Heat stress disorders (adapted from Spain, Ewing, and Clay 1985).

	Symptoms	Causes	Treatments
Transient Heat Fatigue	Decrease in productivity, alertness, coordination and vigilance.	Not acclimatized to hot environment.	Graduate adjustment to hot environment.
Heat Rash ("Prickly Heat")	Rash in area of heavy perspiration; discomfort; or temporary disability.	Perspiration not removed from skin; sweat glands inflamed.	Periodic rests in a cool area; showering/bathing; drying skin.
Fainting	Blackout, collapse.	Shortage of oxygen in the brain.	Lay down.
Heat Cramps	Painful spasms of used skeletal muscles.	Loss of salt; large quantities of water consumed quickly.	Adequate salt with meals; salted liquids (unless advised differently by a physician).
Heat Exhaustion	Extreme weakness or fatigue; giddiness; nausea; headache; pale or flushed complexion; body temperature normal or slightly higher; moist skin; in extreme cases vomiting and/or loss of consciousness.	Loss of water and/or salt; loss of blood plasma; strain on the circulatory system.	Rest in cool area; salted liquids (unless advised differently by a physician).
Heat Stroke	Skin is hot, dry and often red or spotted; core temperature is 40°C (105°F) or higher and rising; mental confusion; deliriousness; convulsions; possible unconsciousness. Death or permanent brain damage may result unless treated immediately.	Thermo-regulatory system breaks down under stress and sweating stops. The body's ability to remove excess heat is almost eliminated.	Remove to cool area; soak clothing with cold water; fan body; call physician/ambulance immediately.

SUMMARY OF THE BODY'S THERMOREGULATORY ACTIONS

While working in a hot or cold environment, the primary purpose of the human thermoregulatory system is to keep the body core temperature (energy content) within narrow limits. For this, a suitable temperature differential (energy flow) must be established between the deep body tissues and the skin. To maintain the core temperature, the body uses a number of procedures:

> To *achieve heat gain,* such as in a cold environment, metabolism is increased through contractions of skeletal muscle.
> To *achieve heat loss,* blood supply to the skin and sweat production are increased.
> To *prevent heat gain,* muscular activities and metabolic functions are reduced.
> To *prevent heat loss,* blood supply to the skin is reduced and muscle contractions are increased.

VARIABILITY OF HEAT GENERATION AND HEAT DISTRIBUTION IN THE BODY

All metabolic functions of the body at rest finally result in the generation of heat. Assuming no interaction with the environment for the moment, the basal metabolism generates approximately 100 J/s. More heat is generated with increased activity. For example, a highly trained 70-kg bicyclist may have an oxygen capacity of 5 L/min. With an energy efficiency of 20%, 80% of the oxygen consumed serves to develop heat that must be dissipated; in this case, approximately 1.4 kJ/s are generated. With no heat exchange, the body temperature would increase quickly. Assume an average heat capacity of the body of 3.4 kJ/kg°C. (This means that 3.4 kJ are needed to increase the temperature of 1 kg of tissue by 1°C.) If there were no heat exchange with the environs, the exemplary bicyclist would increase the total body temperature by nearly 1.5°C in 1 hr if resting, while this same temperature increase would be achieved within about 3 min when bicycling hard.

Of course, this example is highly simplified. In reality, only the core temperature is kept constant while the temperature of the body shell can be varied. Assuming that the shell mass is about one-third of the total body mass and that its temperature may be increased by 6°C without ill effects, the total body energy could be increased by approximately 500 kJ without affecting the core temperature.

The temperature of the shell may be very different from the core temperature. Under cold conditions, with a core temperature maintained at approximately 37°C the trunk may have skin temperatures of about 36°C; the thighs may be at 34°C, with the upper arms and knees at 32°C and the lower arms at 28°C, while the toes and fingers may be at 25°C. These large differences in skin temperatures indicate the effectiveness of the human thermoregulatory mechanisms, particularly shunting and vasoconstriction, in the cold. They also indicate how the body must be protected by clothing to avoid the chilling of body segments below an acceptable temperature. Obviously, fingers and feet need special protection in cold conditions.

The temperatures of the neck and head do not vary much. Here, the "core" is close to the shell, and very little volume between skin and core is available for vasomotor and sudomotor functions. To keep the core warm, skin temperature must be maintained at a rather constant and high level. Hence, in the cold, it is of particular importance to provide external insulation through suitable clothing for head and neck; otherwise, much energy will be lost through the exposed surfaces. In a hot environment, cooling of the frontal part of head is more effective than cooling of other parts of the head (Katsuura et al., 1996).

In a hot environment, the temperature throughout the body is much more constant. The core temperature is approximated in many of the external layers, with the largest differences again found in the extremities, such as the hands and feet.

Acclimatization

Continuous or repeated exposure to hot or cold conditions brings about a gradual adjustment of body functions, resulting in a better tolerance for the climatic stress and in maintenance or improvement of physical work capabilities.

Acclimatization to heat is demonstrated by an increased sweat production, lowered skin and core temperature, and a reduced heart rate, compared with the reactions of the unacclimatized person at first exposure to the hot climate. The process of acclimation is very pronounced within about a week, and full acclimatization is achieved within about two weeks. Interruption of heat exposure for just a few days reduces the lingering effects of acclimatization, which is entirely lost about 2 weeks after returning to a moderate climate.

Heat acclimatization is brought about by improved control of the vascular flow, by an augmented stroke volume accompanied by reduced heart rate, and by higher sweat production. The improvement in sudomotor action is most prominent and manifests itself not only by larger sweat volume but also by an equalization of the sweat production over time and an increase in the activities of the sweat glands of the trunk and the extremities. Perspiration on the face and the feeling of "sweating" become less with heat acclimation, although total sweat production may be doubled after several days of exposure to the hot environment. More volume and better regulation of sweat distribution are the primary means the human body has to bring about dissipation of metabolic heat.

Sudomotor regulation is accompanied by, and intertwined with, vasomotor improvements. The reduced skin temperature (lowered through sweating) allows a redistribution of the blood flow away from the skin surfaces, which need more blood during initial exposure to heat. Acclimation reestablishes normal blood distribution within a week or two. Cardiac output must remain rather constant, even during initial heat exposure, when an increase in heart rate and a reduction in stroke volume occur. Both rate and volume are reciprocally adjusted during acclimation since arterial blood pressure remains essentially unaltered. There may also be a (relatively small) change in total blood volume during acclimation, particularly an increase of plasma volume during the first phase of adjustment to heat.

A healthy and well-trained person acclimates more easily then somebody in poor condition, but training cannot replace acclimatization. If strenuous physical work must be performed in a hot climate, then such work should be part of the acclimation period.

Adjustment to heat will take place whether the climate is hot and dry or hot and humid. Acclimatization seems to be unaffected by the type of work performed, that is, heavy and short or moderate but continuous. It is important that, during acclimation and throughout heat exposure, fluid and salt losses be replaced.

Acclimatization to cold is much less pronounced; in fact, there is doubt that true physiological adjustment to moderate cold takes place when appropriate clothing is worn. The first reaction of the body exposed to cold temperature is shivering—the generation of metabolic heat to counteract heat loss. Also, some changes in local blood flow are apparent. In laboratory exposure to extreme cold conditions, with little shelter offered by clothing, even hormonal and other changes have been observed. Normally, however, the adjustment to cold conditions is more one of proper clothing and work behavior than of pronounced changes in physiological and regulatory functions. Thus, with relatively few changes in physiological functions, food intake or rate of heat production is not much, if at all, changed under "normal" working conditions in "normally cold" temperatures. It appears that, with proper clothing, the actual cold exposure of the body is not very severe and does not require appreciable increases in metabolism or other major adjustments in vasomotor or sudomotor systems. However, there are so-called local acclimatizations, particularly increased blood flow through the hands or in the face.

On average, compared to the male, the female has smaller body mass (about 80%), that is, a smaller heat "sink" than the male; women usually have relatively more body fat and accordingly less lean body mass than men. Their surface area is smaller, however, and their blood volume is smaller as well. Under heat stress, many females show somewhat lower metabolic heat production, have a higher set point, begin to sweat at higher temperatures, and may acclimate more slowly to very hot conditions. Under cold stress, females have slightly colder temperatures at their (thinner) extremities but show no difference in core temperature. Altogether, there are no great differences between females and males with respect to their ability to adapt to either hot or cold climates, with women possibly at a slightly higher risk for heat exhaustion and collapse and for cold injuries to extremities. However, these slight statistical tendencies can be easily counteracted by ergonomic means and may not be obvious at all when only a few persons of either gender are observed.

STRENUOUS WORK IN COLD AND HEAT

Cold and hot climatic conditions (as well as air pollution and high altitude, which are discussed by Kroemer, 1991) variously affect human abilities to perform short or long, moderate or heavy work. The following text is a synopsis of the known effects for use by engineers and managers.

Effects of Cold

As in a hot climate, the body must maintain its core temperature near 37°C in a cold environment. When exposed to cold, the human body first responds by peripheral vasoconstriction, which lowers skin temperature, to decrease heat loss through the skin. Such reduction in blood flow occurs in all exposed areas of the body with the exception of the head, where up to 25% of the total heat loss can take place. If control of blood flow away from the periphery is insufficient to prevent heat loss, shivering sets in. Shivering is a regular muscular contraction mechanism but one that generates no external work since all energy is converted to heat. Muscular activities of shivering and of physical work require increased oxygen uptake, which is associated with increased cardiac output.

Cardiovascular Effects

The necessary increase in cardiac output is brought about mostly by increasing the stroke volume, while heart rate remains at low levels. (An explanation for the increased stroke volume has been found in higher catecholamine levels and in the shifting of blood volume from the periphery into more central circulation which is associated with heightened blood pressure.) Yet, keeping the heart rate low as a reaction to cold exposure opposes the response associated with physical exercise, which is to increase the heart rate to help enlarge cardiac output.

Effects on Body Temperature

The two opposing cardiac responses to cold and exercise affect body temperature. At light work in the cold, core temperature tends to fall after about 1 hr of activity. Cold sensations in the skin regularly initiate reactions leading to lowered skin temperature, yet areas over active muscles can remain warmer because of the heat generated by muscle metabolism. Thus, in the cold, relatively much heat is lost through convection (and evaporation). Which of the opposing physiological cold responses predominates depends on the special conditions, that is, on ambient temperature, type of body activity, and clothing insulation.

While the coldness of air can be felt in the upper respiratory tract, the warming efficiency of the upper respiratory passages is sufficient to preclude cold injuries to lung tissues under normal conditions. Discomfort and constriction of airways may be felt when very cold air is inspired through the mouth. Yet, air temperature is hardly ever too cold for exercise and physical work.

Effects on Energy Cost

Submaximal work in the cold consumes more oxygen than the same work at normal temperatures. (Some of this increased oxygen cost at low work levels is due to shivering.) At higher exercise intensities, oxygen cost in the cold is about the same as at normal temperatures. However, an extra effort is required to "work against" heavy clothing worn to insulate against heat loss.

Regarding maximal exercise levels, fairly little experimental work has been performed. The limited available information indicates that at maximal work levels, a cold climate does not affect the ability for maximal exercise, as long as the exposure does not exceed about 5 hr. In this case, the physiological stimuli provoked by exercise appear to override those of cold. However, if core temperature gets lower, maximal work capacity is reduced, mostly, it would seem, by suppressing heart rate and thus reducing the transport of oxygen to the working muscles in the bloodstream.

Little is known about the effects of cold exposure on endurance. However, a decrease in muscle temperature affects muscle contraction capability negatively, inducing early onset of fatigue.

Dehydration

Dehydration occurs surprisingly easily in the cold, partly because sweating is increased in response to the increased energy demands of working in the cold and because the thirst sensation is suppressed. Also, urine production is increased in the cold, which can trigger water loss through more frequent urination. While the dryness of cold air may cause respiratory irritation and discomfort, severe dehydration through the lungs does not occur since exhaled air is cooled on its way out to nearly the temperature of the inhaled air, returning water vapor by condensation onto the surface of the airways. (This explains the common experience of a "runny nose" in the cold.)

Effects on Mental Performance and Dexterity

If the core temperature of the body drops below about 36°C, vigilance is reduced. Central nervous system coordination suffers at about 35°C, apathy sets in, and loss of consciousness occurs near 32°C. While muscle spindles are initially more active as muscle temperature drops, at about 27°C, their activity is reduced 50% and is completely abolished at about 15°C. In the hands, joint temperatures below 24°C and nerve temperatures below 20°C severely reduce the ability for fine motor tasks. Manual dexterity is reduced as finger skin temperatures fall below 15°C, with tactile sensitivity reduced below 10°C. A nervous block occurs if nerve temperature falls below 10°C; movement becomes impossible and motor skills are completely lost. At about 5°C, skin receptors for pressure and touch cease to function and the skin feels numb (Heus, Daanen, and Havenith, 1995). Frostbite is the result of ice crystals destroying tissue cells.

Working in the Cold: Summary

Strong isometric muscle exertions are impaired only if the muscles are cold. The ability to do light work is reduced in the cold. Endurance activities are impaired only if core or muscle temperatures are lowered and if dehydration occurs. Clothing worn for insulation may hinder work. Dexterity and mental performance suffer in extreme colds.

Effects of Heat

When exposed to whole-body heating, the human body must maintain its "core" temperature near 37°C. It does so by raising its skin temperature, increasing blood flow to the skin, accelerating heart rate, and enlarging cardiac output. The change in blood routing reduces the blood that can be supplied to muscles and internal organs. Yet, if muscles must work, their raised metabolism poses increased demands on the cardiovascular system.

Cardiovascular Effects

The pumping capacity of the heart is between about 25 L/min for "average" adults and 40 L/min for elite athletes. The blood vessels in skin and internal organs can accept up to 10 L and all muscles together up to 70 L/min. Since the available cardiac output is half or less of these 80 L, the ability of the heart to pump blood is the limiting factor for muscular work in a hot climate.

Effects on Muscles

An increase in muscle temperature above normal does not affect the maximal isometric contraction capability of muscle tissue; but the power output of muscles is reduced at higher (and lower) temperatures. Muscle overheating accelerates the metabolic rate, which can make the muscle ineffective if it must work over some period of time. The loss of power and endurance owing to excessive muscle temperature can be counteracted by lowering muscle temperature before exercise. This reduces the cardiovascular strain and blood lactic acid concentration and depletes muscle glycogen at a lower rate.

Dehydration

When working in a hot environment, the body loses water; that is, it gets dehydrated. The body can adapt to heat but not to dehydration. Acute water loss incurred in a short time (in a few hours or less), called *hypohydration,* does not reduce isometric muscle strength (or reaction times) if the water loss is less than 5% body weight. However, fast and large water loss (such as that introduced by diuretics) generates the risk of heat exhaustion, which is primarily the result of fluid volume depletion. Dehydration reduces the body's capacity to perform work of the aerobic or endurance type.

To counteract water loss, one must drink fluid. Plain water is best. If strenuous activities last longer than 1 or 2 hr, diluted sugar additives may help to postpone the development of fatigue by reducing muscle glycogen utilization and improving fluid-electrolyte absorption in the small intestine. Regular, liberally salted food during meals (as customary in the United States) is normally sufficient to counteract salt loss. In fact, salt tablets have been shown to generate stomach upset, nausea, or vomiting in up to 20% of all the athletes who took them.

Effects on Mental Performance

It is difficult to evaluate the effects of heat (or cold) on mental or intellectual performance because of large subjective variations and a lack of practical yet objective testing methods (Ramsey, 1995). However, as a rule, mental performance deteriorates with rising room temperatures, starting at about 25°C for the unacclimatized person; that threshold increases to 30°C or even 35°C if the individual is acclimatized to heat. Brain functions are particularly vulnerable to heat; keeping the head cool improves the tolerance to elevated deep body temperature. A high level of motivation may also counteract some of the detrimental effects of heat. Thus, in laboratory tests of perceptual motor tasks, onset of performance decrement can occur in the low 30°C WBGT range, while very simple mental performance is often not significantly affected by heat as high as 40°C (104°F) WBGT.

Working in the Heat: Summary

Short-term maximal muscle strength exertion is not affected by heat or water loss. The ability to perform high-intensity endurance-type physical work is severely reduced during acclimation to heat, which normally takes up to two weeks. Even after acclimatization is achieved, the demands on the cardiovascular system for heat dissipation and for blood supply to the muscles continue to compete. The body prefers heat dissipation, with a proportional reduction in performance capability. Dehydration further reduces the ability of the body to work; hypohydration poses acute health risks. Mental performance is usually not affected by heat as high as 40°C WBGT.

DESIGNING THE THERMAL ENVIRONMENT

There are many ways to generate a thermal environment that is suitable to the physiological functions for the (acclimatized or nonacclimatized) person and that brings about thermal comfort. The technical means to influence the climate must be seen, obviously, in interaction with the work to be performed, with the acclimatization condition of the individuals, with their clothing, and with their psychological inclination either to accept given conditions or to consider them uncomfortable.

The physical conditions of the climate (humidity, air movement, temperatures) influence the cooling or heating of the body via the heat-transfer functions (radiation, convection, conduction, and evaporation). These interactions, which are listed in Table 9-3, must be carefully considered when designing and controlling the environment.

Microclimate and Thermal Comfort

What is of importance to the individual is not the climate in general, the so-called macroclimate, but the climatic conditions with which it interacts directly. Every person prefers a microclimate that feels "comfortable" under given conditions of adaptation, clothing, and work. The suitable microclimate is highly individual and also variable. It depends on gender, as just discussed. It also depends somewhat on age: with increasing years, the muscle tonus is reduced; older persons tend to be less active, to have weaker muscles, to have reduced caloric intake, and to start sweating at higher skin temperatures. It depends on the surface-to-volume ratio which, in children, for example, is much larger than in adults, and on the fat-to-lean body mass ratio.

Thermal comfort depends largely on the type and intensity of work performed. Physical work in the cold may lead to increased heat production and,

Table 9-3.   Designing the thermal environment to increase (+) or decrease (−) body heat content by changing climate parameters.

Heat Transfer	Air Humidity		Air Movement		Temperatures (as compared to skin) of					
					Air, Water		Solids		Opposing Surface	
	Dry	Moist	Fast	Slow	Hotter	Colder	Hotter	Colder	Hotter	Colder
Radiative	No direct effect		No direct effect		NA		NA		+	−
Convective	No direct effect		−	(−)	+	−	NA		NA	
Conductive	NA		NA		NA		+	−	NA	
Evaporative	−	(−)	−	(−)	−	(−)	NA		NA	

Parentheses indicate that the heat loss is not as pronounced as in the corresponding condition.

hence, to less sensitivity to the cold environment whereas, in the heat, hard physical work may be highly detrimental to the achievement of an energy balance. The effects of the microclimate on mental work are unclear; the only sure commonsense statement is that extreme climates hinder mental work.

Clothing can affect the individual microclimate strongly. The insulating value of clothing is defined in clo units, with 1 clo = 0.115 m^2 °C W^{-1}, which is the insulating value (the reciprocal of thermal conductivity) of the "normal" clothing worn by a sitting subject at rest in a room at about 21°C and 50% relative humidity. Note that this expression assumes that the clothing covers the whole body surface (described by the m^2) with the same insulating value. Air bubbles contained in the clothing material or between clothing layers provide increased insulation, both against hot and cold environments. Permeability to fluid (sweat) and air plays a major role (Lotens, Van de Linde, and Havenith, 1995; Nielsen, Gavhed, and Nillson, 1989; Parsons, 1993, 1995). Clothing colors are important in a heat-radiating environment, such as in sunshine, with darker colors absorbing heat radiation and light ones reflecting incident energy.

Clothing also determines the surface area of exposed skin. More exposed surface areas allow better dissipation of heat in a hot environment but can lead to excessive cooling in the cold. Fingers and toes need special protection in cold conditions because they are so far away from the warm body core. Head and neck are close to the core and have warm surfaces, which is advantageous in the heat but not in the cold.

Convection heat loss is increased if the air moves more swiftly along exposed skin surfaces. Therefore, with increased air velocity, body cooling becomes more pronounced. During World War II, experiments were performed on the effects of ambient temperature and air movement on the cooling of water. These physical effects were also assessed psychophysically in terms of the *wind chill* sensation at exposed human skin. Table 9-4 shows the energy loss depending on air temperature and movement but not considering humidity.

Table 9-5 shows how exposed human skin reacts to energy losses brought about by air velocity at various air temperatures. It lists the wind chill temperature equivalents in degrees Celsius, which reflect the effects of air velocities at various temperatures. Note that these wind chill temperatures are based on the cooling of exposed body surfaces, not on the cooling of a clothed person. Also, these numbers do not take into account air humidity. Under humid conditions, freezing of flesh may occur at wind chill values as low as 3500 kJ/m^{2}hr (800 kcal/m^{2}hr).

Obviously, thermocomfort is also affected by acclimatization, that is, the adjustment of the body and mind to changed environmental conditions. After two weeks, a climate that was rather uncomfortable and that restricted one's ability to perform physical work during the first day of exposure may be quite agreeable. Relatedly, seasonal changes in climate, usual work, clothing, and attitude play a major role in what is regarded as acceptable or not. In the summer, most people are willing to accept as comfortable warmer, more humid, and draftier conditions than they would tolerate in the winter.

Various combinations of climate factors (temperature, humidity, air movement) can subjectively appear similar. The WBGT discussed earlier is one attempt to establish climate factors that, combined, have equivalent effects on people. Similar approaches have been proposed throughout many decades, resulting in about two dozen different scales of *effective temperatures* (Parsons, 1995).

Table 9-4. Wind chill values (energy loss through exposed skin) and their psychological correlates (adapted from U.S. Army MIL-HDBK 759A, 1981).

Approximate Wind Chill Value		Human Sensation
kJ m^{-2}hr^{-1}	kcal m^{-2}hr^{-1}	
200	50	Hot
450	110	Warm
850	210	Pleasant
1600	400	Cool
2500	600	Very cool
3500	800	Cold
4200	1000	Very cold
4900	1200	Bitterly cold

Table 9-5. "Wind chill temperature" for naked skin depending on air temperature and air movement (adapted from U.S. Army MIL-HDBK 759A, 1981).

Wind Speed (m s⁻¹) Calm	Actual Air Temperature (deg C)														
	4	2	-1	-4	-7	-9	-12	-15	-18	-23	-29	-34	-40	-46	-51
2.2	2	-1	-4	-7	-9	-12	-15	-18	-20	-26	-32	-37	-43	-48	-57
4.5	-1	-7	-9	-12	-15	-18	-23	-26	-29	-37	-43	-51	-57	-62	-71
6.7	-4	-9	-12	-18	-21	-23	-29	-32	-34	-43	-51	-57	-65	-73	-79
8.9	-7	-12	-15	-18	-23	-26	-32	-34	-37	-45	-54	-62	-71	-79	-84
11.2	-9	-12	-18	-21	-26	-29	-34	-37	-43	-51	-59	-68	-76	-84	-93
13.4	-12	-15	-18	-23	-29	-32	-34	-40	-46	-54	-62	-71	-79	-87	-96
15.6	-12	-15	-21	-23	-29	-34	-37	-40	-46	-54	-62	-73	-82	-90	-98
17.9	-12	-18	-21	-26	-29	-34	-37	-43	-48	-56	-65	-73	-82	-90	-101

Winds above 18 m s⁻¹ have little *additional* effect

Little Danger

Increasing Danger: Flesh may freeze within 1 min

Great Danger: Flesh may freeze within 30 s

The model shown in Figure 9-2 is typical and rather well accepted. This thermal index does not include radiant heat transfer but reflects how combinations of dry air temperature, humidity, and air movement affect people wearing indoor clothing and doing sedentary or light muscular work. The result is numerically equal to the temperature of still, saturated air, which induces the same sensation. For example, the dashed line in Figure 9-2 indicates that the combination of a dry temperature of about 24°C (A) and a wet temperature of about 17°C (B) generates an effective temperature of about 21°C at low air speed (6 m/min), while strong winds (such as 200 m/min) lower the sensation to that of nearly 17°C effective temperature (ET).

Given these many variables, it is not surprising that the same effective temperature is considered by some people to be too warm and by others as too cold, as graphed in Figure 9-3. At truly cold temperatures, however, much agreement on "cold" is expected while, under very hot conditions, most people will say "hot."

With appropriate clothing and at light work, comfortable ranges of effective temperature (ET) are from about 21° to 27°C ET in a warm climate or during the summer and from 18° to 24°C ET in a cool climate or during the winter. In terms

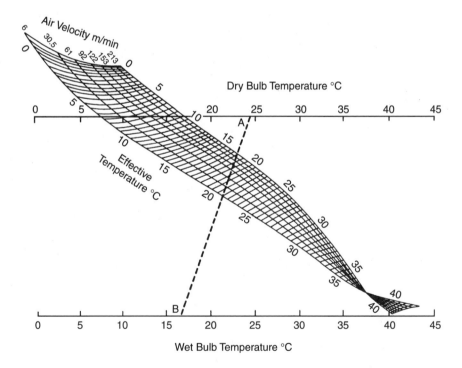

Figure 9-2. Nomogram for deriving the effective temperature from dry and wet bulb temperatures and from air velocity. (U.S. Army, MIL-HDBK-759A, 1981).

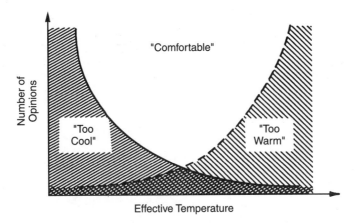

Figure 9-3.   Opinions about effective temperatures.

of body measurements, skin temperatures in the range of 32° to 36°C are considered comfortable, associated with core temperatures between 36.7° and 37.1°C. Preferred ranges of relative humidity are between 30% and 70%. Deviations from these zones are uncomfortable or even intolerable, as Figure 9-4 indicates. Air temperatures at floor and head levels should not differ by more than about 6°C. Differences in temperatures between body surfaces and sidewalls should not exceed approximately 10°C. Air velocity should not exceed 0.5 m/s and, at best, should remain below 0.1 m/s. Further information for the built environment, such as offices, is contained in the ANSI-ASHRAE Standard 55, latest edition.

The temperatures of object surfaces that will be touched with the bare hands should be kept below those listed in Table 9-6.

SUMMARY

The body must maintain a core temperature near 37°C with little variation despite major changes in internally developed energy (heat), in external work performed, in heat energy received from a hot environment, or in heat energy lost to a cold environment.

Heat energy may be gained from, or lost to, the environment by

- Radiation
- Convection
- Conduction
- Evaporation

The major avenues of the body to control heat transfer between core and skin are the efferent motor pathways (muscle tonus), sudomotor pathways (sweat production), and vasomotor pathways (control of blood flow).

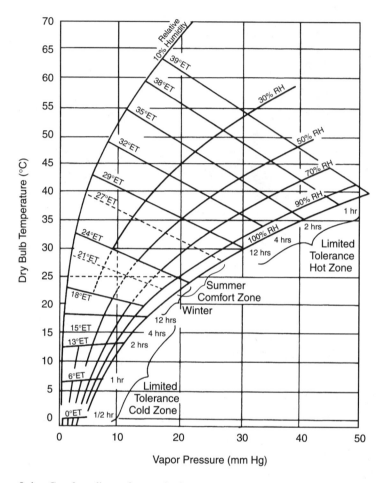

Figure 9-4. Comfort, discomfort, and tolerance zones (U.S. Army, MIL-HDBK-759A, 1981) for young, healthy soldiers, expressed in degrees Celsius of effective temperature ET. The persons are lightly clad, sit, and perform light physical activities. The air moves at no more than 6 m/min in comfort zones and at 60 m/min in tolerance zones.

Muscular activities are the major means to control heat generation in the body. Blood flow control affects heat transfer between body core and skin.

In a hot environment, the body tries to keep the skin hot to prevent heat gain and to achieve heat loss. Sweating is the ultimate means to cool the body surface.

In a cold environment, the body tries to keep the skin cold to avoid heat loss.

Acclimatization to a hot environment includes controlled blood flow to the skin, facilitated sweating, and increased stroke volume of the heart without in-

Table 9-6. Maximal surface temperature (°C) that can be tolerated by human skin without burn risk.

Material	1 sec	4 sec	1 min	10 min	8 hr
Metals					
Uncoated, smooth surface	65	60 ⎫			
Uncoated, rough surface	70	65 ⎬ 50			
Coated with varnish, 50 μm thick	75	65 ⎭			
Concrete, ceramics	80	70 ⎫		all	all
Glazed ceramics (tiles)	80	75 ⎬ 55		48	43
Glass, porcelain	85	75 ⎭			
Plastics					
Polyamid with glass fibers	85	75 ⎫			
Duroplast with fibers	95	85 ⎬ 60			
Teflon, Plexiglas	NA	85 ⎭			
Wood	115*	95	60		
Water	65	60	50		

for contact times of

*Up to 25°C higher for very dry and very light woods.

Adapted from Siekmann, 1990.

crease in heart rate. Acclimitization can be accomplished (in healthy and fit persons) in one or two weeks and be lost just as quickly.

Whether any truly physiological acclimatization to a moderately cold environment takes place is questionable since most of the adjustments made concern proper clothing, within which the body performs at its usual microclimate. However, blood flow to exposed surfaces and to the hands and feet is locally adapted.

The thermal environment is determined by combinations of:

' Air humidity (mostly affecting evaporation)
' Air temperature (affecting convection and evaporation)
' Air movement (affecting convection and evaporation)
' Temperature of solids in touch with the body (affecting conduction)
' Temperature of surfaces distant from the body (affecting radiation)

The combined effects of all or some of these physical climate factors can be expressed in the form of a climate index. Various indices are in use, such as the WBGT and ET scales.

Certain ranges of humidity, temperatures, and air velocity have been identified as "comfortable" for given tasks and clothing.

REFERENCES

ASHRAE (ed.) (1992a). *Thermal Environmental Conditions for Human Occupancy.* ANSI-ASHRAE Standard 55-1992. Atlanta, GA: American Society of Heating, Refrigerating, and Air-Conditioning Engineers.

ASHRAE (ed.) (1992b). 1992 ASHRAE Handbook—*Fundamentals.* Atlanta, GA: American Society of Heating, Refrigerating, and Air-Conditioning Engineers.

Eissing, G. (1995). Climate Assessment Indices. *Ergonomics* 38(1):47–57.

Heus, R., Daanen, H.A.M., and Havenith, G. (1995). Physiological Criteria for Functioning of Hands in the Cold. *Applied Ergonomics* 26(1):5–13.

Hoffman, R.G., and Pozos, R.S. (1989). Experimental Hypothermia and Cold Perception. *Aviation, Space and Environmental Medicine* 66:964–969.

ISO Standard 7243 (1989). Hot Environments. Geneva, Switzerland: International Organization for Standardization.

Katsuura, T., Tomioka, K, Harada, H., Iwanaga, K., and Kikuchi, Y. (1996). Effects of Cooling Portions of the Head on Human Thermoregulatory Response. *Applied Human Science* 15(2):67–74.

Kroemer, K.H.E. (1991). Working Strenuously in Heat, Cold, Polluted Air and at High Altitude. *Ergonomics* 22(6):385–389.

Lotens, W.A., and Havenith, G. (1995). Effects of Moisture Absorption in Clothing on Human Heat Balance. *Ergonomics* 38(6):1092–1113.

Lotens, W.A., and Pieters, A.M.J. (1995). Transfer of Radiative Heat Through Clothing Ensembles. *Ergonomics* 38(6):1132–1155.

Lotens, W.A., Van de Linde, F.J.G., and Havenith, G. (1995). Effects of Condensation in Clothing on Heat Transfer. *Ergonomics* 38(6):1114–1131.

Mairiaux, P., and Malchaire, J. (1995). Comparison and Validation of Heat Stress Indices in Experimental Studies. *Ergonomics* 32(1):58–72.

Malchaire, J. (1995). Methodology of Investigation of Hot Working Conditions in the Field. *Ergonomics* 38(1):73–85.

Nielsen, R., Gavhed, D., and Nillson, H. (1989). Thermal Function of a Clothing Ensemble During Work: Dependency on Inner Layer Fit. *Ergonomics* 32:1581–1594.

Nadel, E.R. (1984). Energy Exchanges in Water. *Undersea Biomedical Research* 11(4):149–158.

Nadel, E.R., and Horvath, S.M. (1975). Optimal Evaluation of Cold Tolerance in Man. Chapter 6A in S.M. Horvath, S. Kondo, H. Matsui, and H. Yoshimura (eds.). *Comparative Studies on Human Adaptability of Japanese Caucasians, and Japanese Americans,* Vol. 1. Tokyo: Japanese Committee of International Biological Program.

Olesen, B.W., and Madsen, T.L. (1995). Measurement of the Physical Parameters of the Thermal Environment. *Ergonomics* 38(1):138–153.

Parsons, K.C. (1993). *Human Thermal Environments.* London, UK: Taylor & Francis.

Parsons, K.C. (1995). International Heat Stress Standards: A Review. *Ergonomics* 32(1):6–22.

Ramsey, J.D. (1990). Do WBGT Heat Stress Limits Apply to both Physiological and Psychological Responses? In *Proceedings, International Conference on*

Environmental Ergonomics IV. Washington, DC: USAF Office of Scientific Research, pp. 132–133.

Ramsey, J.D. (1995). Task Performance in Heat: A Review. *Ergonomics* 32(1):154–165.

Siekmann, H. (1990). Recommended Maximum Temperatures for Touchable Surfaces. *Ergonomics* 21:69–73.

Spain, W.H., Ewing, W.M., and Clay, E. (1985). Knowledge of Causes, Control Aids, Prevention of Heat Stress. *Occupational Health and Safety* 54:4:27–33.

Stegemann, J. (1984). *Leistungsphysiologie,* 3rd ed. Stuttgart, Germany: Thieme.

Stolwijk, J.A.J. (1980). Partitional Calorimetry in Assessment of Energy Metabolism. Chapter in J.M. Kinney (ed.). *Health and Disease.* Columbus, OH: Ross Laboratories, pp. 21–22.

United States Army. (1981). MIL-HDBK-759A. *Human Factors Engineering Design for Army Material.* Redstone Arsenal, AL: U.S. Army Missile Command.

FURTHER READING

Astrand, P.O., and Rodahl, K. (1986). *Textbook of Work Physiology,* 3rd ed. New York, NY: McGraw-Hill.

Baker, M.A. (ed.) (1987). *Sex Differences in Human Performance.* Chichester, UK: Wiley.

Bernard, T.E. (1995). Thermal Stress. Chapter 12 in Plog (ed.). *Fundamentals of Industrial Hygiene.* IL: National Safety Council, pp. 319–345.

Heus, R., Daanen, H.A.M., and Havenith, G. (1995). Physiological Criteria for Functioning of Hands in the Cold. *Applied Ergonomics* 26(1):5–13.

Eissing, G. (1995). Climate Assessment Indices. *Ergonomics* 38(1):47–57.

Kobrick, J.L., and Fine, B.J. (1983). Climate and Human Performance. In D.J. Osborne and M.M. Gruneberg (eds.). *The Physical Environment at Work.* Chichester, Sussex: Wiley, pp. 69–107.

Kroemer, K.H.E., Kroemer, H.B., and Kroemer-Elbert, K.E. (1994). *Ergonomics: How to Design for Ease and Efficiency.* Englewood Cliffs, NJ: Prentice-Hall.

Mairiaux, P., and Malchaire, J. (1995). Comparison and Validation of Heat Stress Indices in Experimental Studies. *Ergonomics* 32(1):58–72.

Mekjavic, I.B., Banister, E.W., and Morrison, J.B. (eds.) (1988). *Environmental Ergonomics. Sustaining Human Performance in Harsh Environments.* Philadelphia, PA: Taylor & Francis.

Olesen, B.W., and Madsen, T.L. (1995). Measurement of the Physical Parameters of the Thermal Environment. *Ergonomics* 38(1):138–153.

Parsons, K.C. (1993). *Human Thermal Environments.* London, UK: Taylor & Francis.

Parsons, K.C. (1995). International Heat Stress Standards: A Review. *Ergonomics* 32(1):6–22.

Chapter 10

Work Schedules and Body Rhythms

OVERVIEW

The human body changes its physiological functions throughout the 24-hr day. During waking hours, the body is prepared for physical work while, during the night, sleep is normal. Attitudes and behavior also change rhythmically during the day. The diurnal rhythms can be unsettled by imposing a new set of time signals and an unfamiliar activity-rest regimen, such as those associated with shift work schedules. Shift work should be arranged to disturb physiological, psychological, and behavioral rhythms as little as possible to avoid negative health and social effects, as well as reductions in work performance.

THE MODEL

Daily rhythms are systems of temporal programs within the human organism. They should be left intact for continued normal functioning, both physically and psychologically, by selection of suitable work schedules.

INTRODUCTION

The human body follows a set of daily fluctuations, called *circadian rhythms* (from the Latin *circa,* "about," and *dies,* "day"; also called *diurnal* from the Latin *diurnus,* "of the day"). These are regular physiological occurrences, observable for example in body temperature, heart rate, blood pressure, and hormone excretion, as graphed in Figure 10-1.

Daily rhythms are systems of temporal programs within the human organism. They are manifest—validated by observation and experience—and well established under rigorous experimental conditions. They are characterized by their

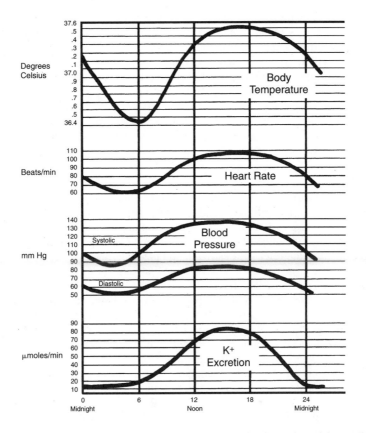

Figure 10-1. Typical variations in body functions over the day (adapted from Minors and Waterhouse, 1981).

persistence under varying external conditions. Each of us is controlled within the body by a self-sustained "internal clock," which runs on a 24- to 25-hr cycle. Several rhythmic programs, such as core temperature, blood pressure, and sleepiness are coupled with each other.

These rhythms can be put out of order, however, by external events, primarily by imposing a new set of external events and time markers on the body. The time marker is called *zeitgeber* (from the German *Zeit,* "time", and *geber,* "giver"). Among the zeitgebers are daily light/darkness, true clocks, and such temporally established activities as office hours and mealtimes. The strengths of these zeitgebers vary.

Human social behavior (the inclination to do certain activities as well as to rest and sleep) follows fairly obvious rhythms and sequences during the day. Of course, we experience other "chronobiological" variations: some are well documented, and others are postulated or mythical, such as dependence on the phases of the moon. So-called biorhythms were a fad a few decades ago: they were said to be regular waves of physiological and psychological events, starting at birth but running in different phases and phase lengths. Whenever "positive" phases of any of these rhythmic phenomena coincided, a person was believed to be under positive conditions, able to perform exceptionally well. In contrast, if "negative" phases concurred, the person was supposedly doing badly. Research has shown conclusively that no such accumulations of positive or negative phases were demonstrable; also, several, or all, of the supposedly existing rhythms were either myths or artifacts (Hunter and Shane, 1979; Persinger, Cooke, and Janes, 1978).

THE MENSTRUAL CYCLE

The female menstrual cycle is regulated through synchronization of the activities of the hypothalamus, pituitary, and ovaries. The typical 28-day time period is usually divided into five phases: (1) preovulatory, or follicular; (2) ovulatory; (3) postovulatory, or luteal; (4) premenstrual; and (5) menstrual. Main hormonal changes occur in the release of estrogen and progesterone around the 21st day of menstruation; estrogen shows a second peak at ovulation. Hormonal release is low during the premenstrual phase.

It has been commonly assumed that hormonal changes during the menstrual cycle have profound effects on a woman's psychological and physiological state. After reviewing the existing scientific work, however, Patkai (1985) found that almost all work has been done to prove correlations between certain behavioral or physiological events and the menstrual phase. But the observable events are fairly weak in their occurrence and, of course, even an existing correlation does not necessarily indicate a causal relationship. Patkai deplores assertions "made by some researchers, on the basis of scientifically poor or scanty data, depicting women as helpless, weak victims of the ebb and flow of their hormones" (p. 88). Certainly, a balanced research position is necessary that considers the close interplay of physiological and psychological factors.

The bulk of existing research relies on self-reported changes in mood and physical complaints in the course of the menstrual cycle, and work performance. While there is neurophysiological evidence that estrogen and progesterone affect brain function, it must be considered that these two hormones have antagonistic effects on the central nervous system, with estrogen stimulating and progesterone inhibiting. Varying hormone production during the menstrual cycle may affect the capacity to perform certain tasks, but the extent to which hormones actually determine performance depends on how a decrease in total capacity may be offset by increased effort. Patkai cites a study in which secretaries showed the highest typing speeds before the onset of menstruation and during the first three menstrual days. The idea of a higher effort on these days was rejected by the secretaries, who considered themselves to be working at full capacity all the time.

The occurrence of negative moods and physical complaints in the majority of women before and during menstruation is fairly well established, but the precise nature of the so-called premenstrual syndrome is not yet determined. According to Patkai, there is evidence that menstruation can bring about negative social behaviors which, however, are mediated by social and psychological factors.

In summary, the hypothesis of reduced performance during the premenstrual and menstrual phases in females is not well supported by objective data.

CIRCADIAN RHYTHMS

The prerequisite for human health is the maintenance of physiological variables in spite of external disturbances. This state of balanced control is called *homeostasis*. But a close look at this supposedly steady state of the body reveals that many physiological functions are, in fact, not constant but show rhythmic variations. Rhythmic variation means that quantitative events (such as ups or downs in body temperature or hormone secretion) follow each other regularly. The periods of different rhythms are quite diverse, such as the heart beating about once every second, body temperature having its peak value every 24 hr, or a menstrual cycle reoccurring every 28 days. While rhythms with a cycle length of 24 hr are called *circadian* or *diurnal* rhythms, those that oscillate faster than once every 24 hr are called *ultradian;* those that repeat less frequently, *infradian.*

Among the circadian rhythms, the best-known physiological variables are body temperature, heart rate, blood pressure, and the excretion of potassium. Most of these variables show a high value during the day and lower values during the night although hormones in the blood tend to be more concentrated during the night, particularly in the early morning hours. The variations during the circadian circle are fairly small, approximately ±1°C for oral temperature; many are in the range of approximately ±15% about the average, such as heart rate and diastolic blood pressure. However, some vary considerably, such as triglycerides, which vary by nearly ±80% in the blood serum, while the sodium content of the urine oscillates even more. The amount by which the variables change during the diurnal variation and the temporal locations of rhythm extremes during the day are quite different among individuals and can change even within one person (Minors and Waterhouse, 1981; Folkard and Monk, 1985).

One way to observe the diurnal rhythms and to assess their effects on performance is simply to observe a person's activities. A person is normally expected to be awake, active, and eating during the day and to be sleeping and fasting at night. Physiological events do not exactly follow that general pattern. For example, body core temperature falls even after a person has been sleeping for several hours; it is usually lowest between 3 and 5 o'clock in the morning. Core temperature then rises quickly when a person gets up. It continues to increase, with some variations, until late in the evening. Thus, body temperature is not a passive response to our regular daily behavior, such as getting up, eating, and working, but is self-governed.

While interactions exist among the external activities and their zeitgebers, the underlying physiological rhythms of the body are solid and self-regulated and remain in existence even if daily activities change. Variations in observed rhythmic events (due to exogenous influences) may occur or even mask the internal regular fluctuations. For example, skin temperature (particularly at the extremities) increases with the onset of sleep, regardless of when this occurs. Turning the lights on increases the activity level of birds, regardless of when this occurs. Thus, skin temperature or activity level does not necessarily indicate the internal rhythm but may, in fact, mask it. Of course, under regular circumstances, there is a well-established phase coincidence between the external activity signs and the internal events. For example, during the night, the low values of physiological functions, such as core temperature and heart rate, are due primarily to the diurnal rhythm of the body; however, they are further helped by nighttime inactivity and fasting. During the day, peak activity usually coincides with high values of the internal functions. Thus, normally, the observed diurnal rhythm is the result of internal (endogenous) and external (exogenous) events that concur. If that balance of concurrent events is disturbed, consequences in health or performance may become apparent.

When a person is completely isolated from external factors (zeitgebers), including regular activities, the internal body rhythms are "running free." This means that the circadian rhythms are free from external time cues and are only internally controlled. Many experiments have consistently shown that circadian rhythms persist when running free, but their time periods are slightly different from the regular 24-hr duration: most rhythms run freely at about 25 hr; some take longer. Since the earth continues to rotate at 24 hr, this experience indicates that body rhythms are independent of external stimuli and follow their own built-in clocks. If a person is subjected again to daily (24-hr) zeitgebers and activities, however, the internal rhythms resume their 24-hr cycles.

Models of Oscillatory Control

The phenomena of human rhythms may be explained by an oscillator model of the human circadian system. This assumes that various overt rhythms are jointly controlled by a few basic oscillators which, however, may have different controlling power.

The basic oscillators are, in turn, controlled by external stimuli and are also influenced by each other. If their intrinsic periods are close together, they synchronize. This internal coordination falls apart, for example, when artificial zeitgebers occur within the entrainment range of one oscillator but outside another. Rhythms controlled by the first oscillator will remain entrained, but those controlled by the other oscillator will begin to run free: internal desynchronization takes place. For example, the sleep/wake cycle may remain entrained, such as at 26 hr, while the temperature rhythm may run freely at a period of 25 hr (Wever, 1985).

To investigate the constancy or temporal isolation of rhythms, two types of experiments have been performed. One uses the absence of any natural or artificial time cues to evaluate the purely internal control. Other experiments use the influence of artificial zeitgebers of various types to evaluate the effects of internal and external factors.

Under constant experimental conditions, that is, without zeitgebers, human circadian rhythms are free-running at about 25-hr periods. This was shown to be true both for isolated individuals and for groups of subjects although some intraindividual, as well as a somewhat smaller interindividual, variability exists (Wever, 1985). Since there is no longer synchronization between the free-running cycles and the 24-hr day, this condition is also called *desynchronization of internal functions* from the 24-hr zeitgebers.

Manipulation of zeitgebers allows laboratory simulation of jet lag or shift work. Experiments with artificial zeitgebers have shown that these play a major role in entraining or synchronizing the internal rhythms so that they follow the periodic time cues. Synchronization of the internal rhythms to time events has been demonstrated to be possible with cycle durations between 23 and 27 hr. (At cues for shorter or longer periods of time, the circadian rhythms are free-running, though often not completely independent of the time cues.) Most researchers have concluded that it is easier to set one's internal clocks "forward," such as in the spring when daylight savings time is introduced in North America and in Europe.

Individual Differences

Experimentation has shown that some people have consistently shorter (or longer) free-running periods than others. For example, those who have short periods are likely to be "morning types," while those with longer internal rhythms are probably "evening types." It appears that females have, on the average, a free-running period about 30 min shorter than that of males. This suggests that females may be more prone to rhythm disorders than males (Wever, 1985).

There is an interaction between aging and circadian rhythms. Rhythm amplitudes are usually reduced with increasing age. This is particularly obvious for body temperature. The temperature rhythm also appears to be advanced relative to the midsleep period, which agrees with the finding of a shift toward morningness with increasing age. Also, if the oscillatory controls lose some of their power with increasing age, as appears to be true, this would indicate a greater susceptibility to rhythm disturbances with increasing age (Kerkhof, 1985).

Daily Performance Rhythms

Given the systematic changes in physiological functions during the day, we expect corresponding changes in mood and performance. Of course, attitudes and work habits are also, and often strongly, affected by the daily organization of getting up, working, eating, relaxing, and going to bed. Experimentally, we can separate the effects of internal circadian rhythms and of external daily organization. For practical purposes, we want to look at the results (e.g., as they affect performance) of the internal and external factors combined.

Early in this century, it was generally held that the morning hours would be best for mental activities, with the afternoon more suitable for motor work. On the other hand, "fatigue" arising from work already performed was believed to reduce performance over the course of the day. For simple mental work, such as recording numbers, it was observed that performance showed a pronounced reduction early in the afternoon. This was labeled the "postlunch dip." However, this reduction in performance was not paralleled by a similar change in physiological functions; for example, body temperature remains fairly unchanged at that period of the day. Hence, it was postulated that the interruption of activities by a noon meal and the following digestive activities of the body would bring about this often observed reduction in performance.

Such postlunch dips are found mostly with activities that can be related to the psychological "arousal" level. Lunchtime might bring about increased lassitude, a status of deactivation, paralleled by, or associated with, increased blood glucose and pulse rate, possibly the results of food ingestion. In this case, the postlunch dip appears to be caused by the exogenous masking effect of food intake rather than by endogenous circadian effects.

In other activities, however, primarily those with medium to heavy physical work, no such dip has been found after lunch, except when the food and beverage ingestion was very heavy and only if true physiological fatigue had been built up during the prelunch activities.

In summary, some of the many different activities performed during the day strongly follow a circadian rhythm and some less strongly. For some people, exogenous masking effects may be more pronounced than for others. For example, information processing in the brain (including immediate or short-term memory demands), mental arithmetic activities, or visual searches may be strongly affected by personality or by the length of the activity and by motivation. Thus, it appears that we cannot make "normative" statements about diurnal performance variations or abilities during regular working hours.

SLEEP

Two millennia ago, Aristotle thought that, during wakefulness, "warm vapors" in the brain built up that needed to be dissipated during sleep. In the 19th century, there were two opposing schools of thought: one that sleep was caused by some "congestion of the brain by blood," the other that blood was "drawn away from the brain." Also, "behavioral" theories were common in the 19th century, such that sleep was the result of an absence of external stimulation or that sleep was not a passive response but an activity designed to avoid fatigue from occurring. Early in the 20th century, it was thought that various sleep-inducing substances accumulated in the brain, an idea taken up again in the 1960s. In the 1930s and 1940s, various "neural inhibition" theories were discussed, including sleep-inducing "centers," such as arousal centers in the reticular formation of the brain (Horne, 1988).

Restorative theories about the function of sleep focus on various types of "recovery" from the wear and tear of wakefulness. Alternative theories reject this idea and claim that sleep is not restorative but simply a form of instinct or nonbehavior to occupy the unproductive hours of darkness; through relative immobility of the body, sleep may be a means to conserve energy (Horne, 1988).

According to Horne, it is convenient to consider that the regulation of alertness, wakefulness, sleepiness, sleep, and of many physiological functions is under the control of two "central clocks" of the body. One controls sleep and wakefulness, the other physiological functions, such as body temperature. Under normal conditions, the internal clocks are linked together so that body temperature and other physiological activities increase during wakefulness and decline during sleep. However, this congruence of the two rhythms may be disturbed, for instance, by night-shift work, where a person must be active during nighttime and sleep during the day. As such patterns continue, the physiological clocks adjust to the external requirements of the new sleep/wake regimen. This means that the formerly well-established physiological rhythm flattens out and, within a period of about two weeks, reestablishes itself according to the new sleep/wake schedule.

Sleep Phases

The brain and muscles are the human organs that show the largest changes from sleep to wakefulness: their activities can be observed and recorded by electrical means.

To observe human sleep, electrodes attached to the surface of the scalp pick up electrical activities of the cortex, which is also called *encephalon* because it wraps around the inner brain. Thus, the measuring technique is named electroencephalography, or EEG. The EEG signals provide information about the activities of the brain. It is also common to record the electrical activities associated with the muscles that move the eyes and those in the chin and neck regions. The electrical recording of muscle activities is called electromyography (from the Greek "*myo*"), or EMG. Interestingly, in sleep research, EMG signals are not analyzed to the extent common in biomechanics or physiology (Basmajian and DeLuca, 1985).

Electroencephalographic signals can be described in terms of amplitude and frequency. The amplitude is measured in microvolts (μV), with the amplitude rising as consciousness falls from alert wakefulness through drowsiness to deep sleep. The frequencies of EEGs are measured in hertz; the frequencies observed in human EEGs range from 0.5 to 25 Hz. Frequencies above 15 Hz are called *fast waves,* and frequencies below 3.5 Hz *slow waves.* Frequency falls as sleep deepens; "slow-wave sleep" (SWS) is of particular interest to sleep researchers.

Certain frequency bands have been given Greek letters. The main divisions are as follows:

- Beta, above 15 Hz. Such fast waves of low amplitude (under 10 μV) occur when the cerebrum is alert or even anxious.
- Alpha, between 8 and 11 Hz. These frequencies occur during relaxed wakefulness, when there is little information input to the eyes, particularly when they are closed.
- Theta, between 3.5 and 7.5 Hz. These frequencies are associated with drowsiness and light sleep.
- Delta, slow waves under 3.5 Hz. These are waves of large amplitude, often over 100 μV, and they occur more often as sleep becomes deeper.

Also, certain occurrences that appear regularly in EEG waves have been labeled, such as vertices, spindles, and complexes; these are associated with sleep characteristics.

The importance placed on EEG and EMG events by sleep researchers has been changing over decades. Currently, EMG outputs of the eye muscles are most often used as the main descriptors: sleep is divided into periods associated with rapid eye movements, REM, and those without, non-REM. Non-REM conditions are further subdivided into four stages according to their associated EEG characteristics. Table 10-1 lists these.

Table 10-1.   Sleep stages (adapted from Horne, 1988; and Rechtschaffen and Kales, 1968).

Condition	Muscle EMG	Brain EEG	Sleep stage	Total sleep time (avg. %)
Awake	Active	Active, alpha & beta	0	—
Drowsy, transitional "light sleep"	Eyelids open and close, eyes roll	Theta, loss of alpha, vertex sharp waves	1, non-REM	5
"True" sleep		Theta, few delta, sleep spindles, K-complexes	2, non-REM	45
Transitional "true sleep"		More delta, SWS (< 3.5 Hz)	3, non-REM	7
Deep "true" sleep		Predominant delta SWS (< 3.5 Hz)	4, non-REM	13
Sleeping	Rapid eye movements, other muscles relaxed	Alert, much dreaming alpha and beta	REM	30

The REM sleep phase is accompanied by irregular breathing and heart rate, and by low-voltage, fast-brain activities visible in the EEG. In the non-REM sleep phase, regular and slow breathing and heart rates occur, and the EEG activity is slow but shows high voltage. These phases change cyclically during sleep and are probably co-organized between diverse brain regions, thus involving two or more oscillators. The REM/non-REM cycles occur in roughly 1.5-hr timings, an ultradian (shorter-than-daily) rhythm. However, this 1.5-hr duration has large within- and between-subjects variability and appears to shorten in the course of a night's sleep, accompanied by a relative lengthening of the REM portion.

In 1963, Kleitman (according to Lavie, 1985) suggested the existence of a basic rest-activity cycle (BRAC) running throughout the day. Such a BRAC would explain the observation that many renal, gastric, eating, and behavioral as well as mental activities follow roughly a 1.5-hr cycle. In particular, it would explain fluctuations in alertness and arousal, inattention, daydreaming, and sleepiness during the waking hours. Thus, Kleitman (according to Lavie) proposed that suitable jobs be organized into working periods and breaks so that each work unit lasts approximately 1.5 hr.

Sleep Loss and Tiredness

If a person does not get the usual amount of sleep, the apparent result is tiredness, and the obvious cure is to get more sleep. Shift workers are particularly subject to this problem, when they have to make up during the day for sleep lost during their regular resting hours. Figure 10-2 shows the effects of sleep loss on body temperature—the temperature is raised during the night and morning but keeps its phase.

It is of some interest to note that it is not entirely clear why humans (or animals) need sleep. There is the general opinion that sleep has recuperative benefits, allowing some sort of restitution or repair of tissue or brain following the "wear and tear" of wakefulness. Certainly, sleep is accompanied by rest and, to a large extent, by energy conservation. But a human can attain similar relaxation during wakefulness when not forced to be active. Regarding restitution of the body, it has been found that more of certain hormones are released during sleep than during wakefulness; prominent among these is the human growth hormone. However, few such "positive" sleep events have been noted. Protein in tissues is continuously broken down into its amino acid building blocks or reconstituted from recent food intake. If such breakdown were excessive during wakefulness, then the rate of synthesis should be especially high during sleep. This was not found to be the case: in fact, during sleep, protein synthesis is low and breakdown increased, leading to the loss of protein through dissolution instead of an increase through restitution.

Many experiments have failed to show restitutive physiological effects of sleep; in fact, even moderate sleep deprivation has little physiological effect (except for clear signs of impairment of the central nervous system), as discussed by Horne (1988). For example, sleep deprivation does not impair muscle restitution or the physiological ability to perform physical work. Apparent reductions in physical exercise capability owing to sleep deprivation (such as those reported by Froeberg, 1985) may be mostly due to reduced psychological motivation rather than to a decrease in physiological capabilities. The effects of sleep deprivation on body functions are not clear but may be less consequential than is often believed.

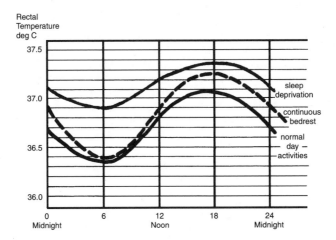

Figure 10-2. Changes in body temperature associated with bedrest, normal activities, and sleep deprivation (schematically from Colligan and Tepas, 1986).

The lack of experimental findings regarding the physical benefits of sleep is somewhat surprising because this does not coincide with common experience. After physically strenuous work or exercise, tired and hurting muscles are obviously recuperated after a good night's sleep. Would physiological restoration processes, such as tissue rebuilding or the metabolizing of lactic acid, which apparently take place during sleep, really occur as well during waking rest?

In contrast, the restorative benefits of sleep to the brain are fairly well researched. Two or more nights of sleep deprivation bring about psychological performance detriments, particularly reduced motivation to perform (but apparently not a reduction in the inherent cognitive capacity), behavioral irritability, suspiciousness, slurred speech, and other performance reductions. However, while these changes indicate some central nervous system (CNS) impairment owing to sleep deprivation (a need for the brain to sleep), Horne (1988) states that they are not as extensive as one might expect for a person considered to need 8 hr of sleep per day for brain restitution. Even though one feels tired, mental performance is still rather normal after up to two days of sleep deprivation on stimulating and motivating tasks; boring tasks, however, show performance reduction. (All task performance is reduced after more than two nights of sleep deprivation.) It is of some interest to note that performance levels are lower during nighttime activities, when the body and brain usually rest.

Horne (1988) speculates that, during normal mental tasks, there is an overcapacity of cerebral neural networks since many are not fully used. During sleep deprivation, however, possibly as a result of missing "restitution", the extra circuits become used, at first without overt effect on performance. With increasing deprivation, more circuits become impaired, so that all available circuits finally become used up and performance drops.

Within the human body, only the brain assumes a physiological state during sleep that is unique to sleep and cannot be attained during wakefulness. While muscles, for example, can rest during relaxed wakefulness, the cerebrum remains in a condition of "quiet readiness," prepared to act on sensory input, without diminution in responsiveness (Horne, 1988). Only during sleep do cerebral functions show marked increases in thresholds of responsiveness to sensory input. In the deep sleep stages associated with slow-wave non-REM sleep, the cerebrum is apparently functionally disconnected from subcortical mechanisms. Regardless of whether the cerebrum needs off-line recovery during sleep from the demands of waking activities or whether it simply disconnects and withdraws, it is undisputed that the brain needs sleep to restitute, a process that cannot take place sufficiently during waking relaxation (Horne, 1988).

Apparently, not all human sleep is essential for brain restitution. For example, after sleep deprivation, usually not all lost sleep is reclaimed; long-term studies with volunteer subjects have shown that a sleep period that is 1 to 2 hr shorter than usual can be endured for many months without any consequences. It seems that the first 5 to 6 hr of regular sleep (which happen to contain most of the slow-wave non-REM sleep and at least half the REM sleep) are obligatory to retain psychological performance at normal level but that more sleep, called facultative or optional, serves mostly to occupy unproductive hours of darkness, with dreams mostly in REM but not totally confined to it the "cinema of the mind" (Horne, 1988, p. 313).

Normal Sleep Requirements

While there are, as usual, variations among individuals, certain age groups in the Western world show rather regular sleeping hours. For example, young adults sleep, on the average, 7.5 hr (with a standard deviation of about 1 hr). Some people are well rested after 6.5 hr of sleep or less, while others habitually take 8.5 hr and more. The amount of slow-wave sleep in both short and long sleepers is about the same, but the amounts of REM and non-REM sleep periods differ considerably. Individuals naturally sleeping less than 3.5 hr are very rare among middle-aged people; no true nonsleepers have ever been found among otherwise healthy persons (Horne, 1988).

Many people who can sleep for just a few hours per day are able to keep up their performance levels even if the attained total sleep time is shorter than normal. The limit seems to lie around 5 hr of sleep per day, with even shorter periods still somewhat useful.

People can learn to extend their sleeping hours, such as taking another hour or more after the alarm rang. The regular sleep duration can also be extended by daytime napping.

In summary, the body's physical need for restitutive sleep is not clearly understood. Some sleep, particularly of the SWS type, may be the only state in which the cerebrum can obtain some form of off-line recovery. No organ other than the brain shows a physiological state during sleep that is unique to sleep and cannot be attained during wakefulness. Thus, it appears that the restorative effects of sleep are mostly beneficial for the brain, less so (and not well understood) for the rest of the body. Moderate loss of sleep is not very consequential for performance.

PROLONGED HOURS OF WORK AND SLEEP DEPRIVATION

There are conditions demanding continuing work for long periods of time, such as a full day or longer. This means not only long periods of work without interruption but also sleep deprivation. Hence, negative results of such long working spells are partly a function of the long work itself and partly of lack of sleep. A discussion of this topic is difficult because different types of tasks may be performed, because motivation can play an important role in work performance, and because wakefulness or sleepiness appears in cycles during the day.

Performing Tasks

Performance of a task that requires uninterrupted periods of a half-hour or longer is more negatively affected by sleep loss than a shorter task. Also, if such a longer task must be replicated, performance is likely to become worse with each successive repetition. Monotonous tasks are strongly affected by sleep deprivation, but a task new to the operator is less affected. On the other hand, a complex task can be more affected than a simple one. Thus, both monotonous and complex tasks should not be required of workers over long periods of time, particularly if sleepiness is a factor.

Tasks that are paced by the work itself deteriorate more with sleepiness than self-paced tasks. Accuracy in performing a job may still be quite good, even after losing sleep, but it takes longer to perform the job. Froeberg (1985) also found that a task that is interesting and appealing, even if it includes complex decision making, can be performed rather well, even over long periods of time. But, if the task is disliked and unappealing, decision making is prolonged. Memory also degrades when people are required to stay awake for long periods of time. Such memory deprivation appears to affect both long-stored memory information and short-term memory.

In summary, performance of "mental tasks" that take more than 30 min deteriorates if the tasks offer little novelty, interest, or incentive, or if they are highly complex. However, performance of tasks that require decision making, problem solving, or concept information that is highly interesting or rewarding is more resistant to deterioration.

Incurring Performance Decrement and Recovering from It

Many performance functions are lowered after one night without sleep. The deterioration becomes more pronounced after two, three, or four nights of sleep deprivation. After missing four nights, very few people are able to stay awake and to perform even if their motivation is very high (Froeberg 1985). The following discussion assumes sleep deprivation of at least one night.

With increasing time at work, so-called microsleeps occur more frequently. The subject falls asleep for a few seconds, but these short periods (even if frequent) do not have much recuperative value because the subject still feels sleepy and performance still degrades. Another commonly observed event during long working times coupled with lack of sleep is periods of no performance, also known as lapses or gaps. These are short periods of reduced arousal or even of light sleep.

Naps lasting 1 to 2 hr improve subsequent performance. However, a person who is awakened from napping during a deep-sleep phase may exhibit "sleep inertia," with low performance; this inertia may last for up to 30 min. When a nap is taken may have a great deal to do with its effectiveness: for example, the common early afternoon nap has surprisingly little effect on performance of subsequent work. On the other hand, naps of at least 2 hrs taken in the late evening or during the night, when the diurnal rhythm is falling, have positive effects lasting several hours, provided that the amount of sleep loss incurred until this moment is moderate, such as one night without sleep (Gillberg, 1985; Rogers et al., 1989).

Scientific findings about the usefulness of naps (for people who missed at least one night's sleep) do not appear to concur with common experience. Many people who take short naps, particularly after lunch, claim that those 5 to 15 min of rest are very helpful, even necessary, to make them "ready and fit" for continued work. Perhaps the recuperative effects are too subtle, too much an interaction between physiological and psychological traits and habit to be easily demonstrated in a scientific experiment.

If long working periods are unavoidable and the work consists largely of mental activities, one may try to have workers perform physical exercises. However, this is not a sure way to prevent deteriorating performance. Also, white noise may improve performance slightly: stirring music may help. Drugs, particularly amphetamines, can restore performance to a normal level when given after even three nights without sleep (Froeberg, 1985).

Recovery from sleep deprivation is quite fast. A full night's sleep, undisturbed, probably lasting several hours longer than usual, restores performance efficiency almost fully.

In summary, task performance is influenced by three factors: the internal diurnal rhythm of the body, the external daily organization of work activities, and subjective motivation and interest in the work. Each factor can govern, influence, or mask the effects of the other factors on task performance. Physical and mental performance capabilities deteriorate during long-continued work accompanied by sleep loss. The performance decrement depends mainly on the task characteristics and on the motivation of the person working. Task duration, monotony, complexity, and repetitiveness have particularly negative effects, while exceptionally high motivation may prevent any performance decrement. Performance is particularly low during "low" periods of the circadian cycle, such as in the early morning hours. Short naps during the work have some beneficial effect on performance. One long night's sleep usually restores performance to a normal level, even after extensive sleep deprivation.

SHIFT WORK

One speaks of shift work when two or more persons, or teams of persons, work in sequence at the same workplace. Often, each worker's shift is repeated, in the same pattern, over a number of days. For the individual, shift work means attending the same workplace regularly at the same time (continual shift work) or at varying times (discontinuous, rotating shift work).

The Development of Shift Work

Shift work is not new. In ancient Rome, it was decreed that deliveries be done at night to relieve street congestion. Bakers have traditionally worked through the late night and early morning hours. Soldiers and firefighters always worked night shifts. Many alternate work systems have been used; for example, farmers used to start work at dawn and stop at dusk. "Peddlers, politicians, professors" and others have always tried to set their own work schedules, which are often highly variable to suit specific circumstances.

With the advent of industrialization, long working days became common, with teams of workers relaying each other to maintain blast furnaces, rolling mills, glass works, and other workplaces where continuous operation was desired. Covering the 24-hr period with either two 12-hr or three 8-hr shifts became common practice. Technological or economical concerns have made shift work generally accepted in many industries, trades, and services, even in developing countries.

Since the Industrial Revolution, when 12-hr shifts were common, drastic changes in work systems have occurred. In the early 20th century, the then common 6-day workweeks with 10-hr shifts were shortened. Today, many work systems use the 8-hr per day/5 workdays per week arrangement, which was first introduced in many countries in the 1960s. In general, the number of days worked per week was reduced, usually to allow two weekend days to be free; also, the number of hours worked per day was reduced. In 1988, the West German secretary of labor proposed a system with a "compressed workweek" of only 4 days a week but 9 hrs per day. In 1996, the average number of hours worked per week was just over 32 in Germany. Many employees in Germany, particularly if working for the government, leave their workplaces early on Friday afternoon to return after a long weekend on Monday morning.

Among other features, many modern industrial work systems require two or more work crews to utilize machinery and keep processes going through much or all of the 24-hr day or the 7-day week. Thus, as seen by the individual worker, there appears to be a trend toward new forms of discontinuous systems, such as weekly rotations of alternating day and night shifts and three-shift systems that cover 24 hr. Irregular rotation appears increasingly, as well as six or seven consecutive similar shifts.

Though estimates vary and depend on the definition of shift work, probably about 25% of all workers in developed countries are on some kind of shift schedule (Monk and Tepas, 1985). Furthermore, working overtime has become a fairly regular feature in many employment situations.

Shift Systems

Shift work is different from "normal" day work in that work is performed regularly during times other than morning and afternoon; and/or at a given workplace, more than that one shift is worked during the 24-hr day. A shift may be shorter or longer than an 8-hr work period.

There are many diverse shift systems. For convenience, they can be classified into several basic patterns, but any given shift system may comprise aspects of several patterns. Kogi (1985) lists four particularly important features of shift systems: (1) whether a shift extends into hours that would normally be spent in sleep; (2) whether it is worked throughout the entire 7-day week or includes days of rest, such as a free weekend; (3) the number of shifts into which the daily work hours are divided, that is, two, three, or more shifts per day; and (4) whether the shift crews rotate or work the same shifts permanently. All these aspects, shown in Figure 10-3, are of particular concern with respect to the welfare of the shift worker, the work performance, and the organizational scheduling.

Other identifiers of shifts and shift patterns are the starting and ending time of a shift; the number of workdays in each week; the hours of work in each week; the number of shift teams; the number of holidays per week or per rotation cycle; the number of consecutive days on the same shift, which may be fixed or variable; and the schedule by which an individual worker either works or has a free day or days (Kogi, 1985).

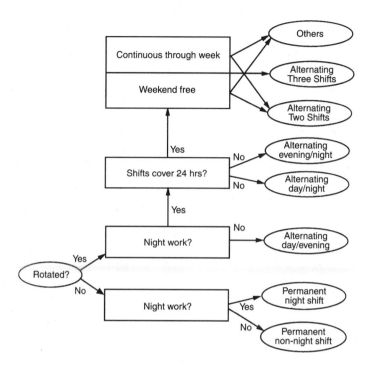

Figure 10-3. Flowchart of key features of shift systems (adapted from Kogi, 1985). Note that other shift attributes are possible.

In terms of organizing the schedule, it is easiest to set up a permanent or weekly rotation schedule. Several such solutions are shown in Table 10-2.

Many examples of work schedules are presented and discussed in the literature. (See, e.g., Colligan, and Tepas, 1986; Colquhoun, 1985; Eastman Kodak Company, 1986; Folkhard and Monk, 1985; Johnson et al., 1981; Knauth, 1996; Tepas and Monk, 1986.) In most systems used today, the same shift is worked for five days, usually followed by two free days during the weekend. This regimen, however, does not cover evenly all 21 shift periods of the week; thus, additional crews are needed to work on weekends or under other "odd" arrangements. If one uses three shifts a day (numbered 1, 2, and 3) with four teams, the shift system (for one team) is 1-1-2-2-3-3-0-0, with a 6-workday 2-free day ratio and a cycle length of 8 days; this is known as the metropolitan rotation. The continental rotation, which also assumes three shifts per day and four crews, has the sequence 1-1-2-2-3-3-3, 0-0-1-1-2-2-2, 3-3-0-0-1-1-1, 2-2-3-3-0-0-0; its workday/free-day ratio is 21/7 and its cycle length exactly 4 weeks.

The ratio of workdays versus free days in a complete cycle is an important characteristic of any shift system. Table 10-3 presents a number of other features that describe different shift systems.

Table 10-2. Examples of 5-workdays-per-week shift systems (adapted from Kogi 1985).

System	Work days/Free days	Shift sequence
Permanent day shift	5/2	1-1-1-1-1-0-0, 1-1-1-1-1-0-0
Permanent evening shift	5/2	2-2-2-2-2-0-0, 2-2-2-2-2-0-0
Permanent night shift	5/2	3-3-3-3-3-0-0, 3-3-3-3-3-0-0
Rotations:		
Alternating day-evening	10/4	1-1-1-1-1-0-0, 2-2-2-2-2-0-0
Alternating day-night	10/4	1-1-1-1-1-0-0, 3-3-3-3-3-0-0
Alternating day-evening-night	15/6	1-1-1-1-1-0-0, 2-2-2-2-2-0-0
		3-3-3-3-3-0-0 (forward rotation)
		or
		1-1-1-1-1-0-0, 3-3-3-3-3-0-0
		2-2-2-2-2-0-0 (backward rotation)

1 represents day shift; 2, evening shift; 3, night shift; 0, free day, i.e., without scheduled shift.

A recent trend is toward *flextime,* a word indicating a flexible arrangement of work hours during the day. This arrangement might be better described as "sliding time" (akin to the German *Gleit-Zeit*) since it allows the employee to distribute the prescribed number of working hours per shift (for example, 8) over a longer block of time (10 hr), as long as a "core" time (of, say, 6 hr) is covered during which all workers must be present. Thus, workers can slide or float the working time across the core time as long as the start of work is at any time before, and the end of work at any time after, the core. Flextime is often, but not necessarily, combined with compressed workweeks, discussed below. Table 10-4 lists potential advantages and disadvantages of flextime.

Table 10-3. Characteristics of shift arrangements (Kogi 1985).

Cycle length	$C = W + F$	
Free days per year	$D = 365\ F\ (W + F)^{-1}$	
Number of days worked before the same set of shifts occur on the same days of the week	$R = W + F$	if $(W + F)$ is a multiple of 7
	$R = 7(W + F)$	if $(W + F)$ is not a multiple of 7

Note: W represents workday; F, free day.

Table 10-4.   Potential advantages and disadvantages of flextime (adapted from Tepas, 1985).

Potential advantages

Generally appealing
Flexible work times with no loss in base pay
Increases day-to-day flexibility for free time
Reduces commuting problems and costs
Workforce size can adjust to short-term fluctuations in demand
Less fatigued workers
Reduces job dissatisfaction/increases job satisfaction
Increases democracy in the workforce
Recognizes and utilizes employee's individual differences
Reduces tardiness
Reduces absenteeism
Reduces employee turnover
Increases production rates
Better opportunites to hire skilled workers in tight labor markets

Potential disadvantages

Irregularity in workhours produced by short-term changes in demand
Difficulty covering some jobs at all required times
Difficulty in scheduling meetings or training sessions
Poorer communication within the organization
Poorer communication with other organizations
Increases energy and maintenance costs
Increases buffer stock for assembly-line operations
More sophisticated planning, organization and control
Reduces quantity or quality of services to the public
Requires special time-recording
Additional supervisory personnel
Extension of health and food service hours

Compressed Workweeks

One speaks of a compressed workweek if the regular number of weekly work hours (say, 40 hr) are compressed into fewer than the usual workdays per week (say, 5 days). This results in more work hours per day, but fewer workdays. For example, the common 40 hr of work per week may be performed in only 4 or even 3 days per week (instead of 5 days). This allows a worker to have 3 or 4 free days each week. Apparently, this is attractive to many employees and employers: There are more work-free days, the number of trips to and from work are reduced, and there are fewer setups and closedowns at work (Hurrell and Colligan, 1985). However, there are concerns about increased fatigue due to long workdays and about reduced performance and safety; see Table 10-5.

Table 10-5.   Potential advantages and disadvantages of compressed workweeks (adapted from Tepas, 1985).

Potential advantages
Generally appealing
Increases possibilities for multi-day off-the-job activities
Reduces commuting problems and costs
More time per day for scheduling meetings or training sessions
Fewer start-up and/or warm-up expenses
Increases production rates
Improves the quantity or quality of services to the public
Better opportunities to hire skilled workers in tight labor markets

Potential disadvantages
Decrements in job performance due to long work hours, or to "moonlighting" on "free" days
Overtime pay required
More fatigued workers
Increases tardiness and "leaving work early"
Increases absenteeism
Increases employee turnover
Increases on-the-job and off-the-job accidents
Decreases production rates
Scheduling problems if the organization operations are longer than the workweek
Difficulty in scheduling child care and family life during the workweek
Contrary to traditional objectives of labor unions
Increases energy and maintenance costs

   The type of work to be performed is a major determiner of whether compressed workweeks can and should be used. Thus, long working days have been used mostly in cases in which persons such as firefighters wait or are on standby for much of the shift. Also, activities that require only few or small physical efforts, that are diverse and interesting yet fall into routines, have been performed in long shifts. Examples are nursing, clerical work, administrative work, technical maintenance, computer supply operations, and supervision of automated processes. Few examples are available in the literature of long shifts that include manufacturing, assembly, machine operations, and other physically intensive jobs. Furthermore, information has been gathered mostly from subjective statements of employees, some limited psychological test batteries, and by scrutinizing performance and safety records in industry.

   The results are contradictory and spotty and apparently depend heavily on the given work conditions. In some reports, production and performance levels are high shortly after introduction of a compressed workweek but fall off after prolonged periods on such a schedule. However, other observations have not shown this trend. The people involved often indicate significantly increased satisfaction, which may be due more to longer blocks of leisure time than to improvements in the work (Lateck and Foster, 1985). In one experimental field study, health employees changed from a 5-day/40-hr workweek to a 4-day/40-hr schedule for four months and then returned to the original schedule. In general, the effectiveness of the organization was somewhat improved by the compressed schedule; employees were more satisfied and reported that their personal lives had benefited. However, these positive effects were insufficient to warrant continuing the trial. The return to the original 5-day workweek was not welcomed by the participants (Dunham, Pierce, and Castaneda, 1987).

Working very long shifts, such as 12 hr, is likely to introduce drowsiness and some reduction in cognitive abilities, motor skills, and general performance during the course of each shift as the workweek progresses. There appears to be a potential for fatigued workers to take careless shortcuts to complete a job, and for work practices to be less safe in tasks "that are tedious because of high cognitive or information-processing demands, or those with extensive repetition" (Rosa and Colligan, 1988, p. 315).

WHICH SHIFT SYSTEMS ARE SUITABLE?

Human beings are used to daylight activity, with the night reserved for rest. This appears to be an inherent feature, as it is governed by the internal clocks of diurnal rhythms. Night work, then, appears to be "unnatural;" however, this does not necessarily mean that it is harmful. But working at night seems to generate "stress," which may be light or severe, depending on the circumstances and the person.

Organizational criteria by which to judge the suitability of shift systems include the number of shifts per day, the length of every shift, or the times of the day during which there is no work done; the coverage of the week by shifts; and shiftwork on holidays. These "independent variables" have been discussed earlier.

Among the "dependent variables" is the performance of workers on shift schedules. Is the same output to be expected, regardless of the time of work during the 24-hr day? Are specific activities better performed at certain shifts? Does change in shift work scheduling affect the shift worker's output? Is work on certain shifts more likely to show accidents?

Another "dependent variable" is the health of the shift worker. Do certain shift regimes affect physiological or psychological well-being? For example, does the inability to sleep at night, when a night shift needs to be worked, affect the worker's health? What is the effect of shift work on social interactions with family, friends, and society in general?

Thus, several factors should be considered in selecting and scheduling shift work: health and well-being, performance and accidents, psychological and social aspects. All interact with each other but not always in the same direction. Conclusions for selecting a suitable work regimen depend on the given conditions.

Health and Well-Being

It is obvious that a strong circadian system exists within the body that is remarkably resistant to sudden large changes in routine. Given the stubbornness of this internal system, theoretical findings, common sense, and personal experiences suggest that the normal synchrony of behavior in terms of rest and activity sequences should be maintained as much as possible. Thus, work schedules should be arranged in accordance with the internal system or, if this is impossible (for example, if night work is necessary), to disturb the internal cycles as little as possible. One logical conclusion from this consideration is that work activities that run counter to internal rhythms should be kept as short as feasible so that workers can return to the normal cycle as quickly as possible. For example, one should schedule single night shifts interspersed by normal workdays instead of requiring a worker to do a series of night shifts, as shown in Table 10-2. Such a series of night shifts upsets the internal clock, while a single night shift will not disturb the entrained cycle severely.

The other solution, both theoretically sound and supported by experience, is to entrain new diurnal rhythms. It takes regular and strong zeitgebers to overpower the regular signals, such as light and darkness. For shift work, this means that the same setup (such as working the night shift) should be maintained for long periods of time (several weeks, even months) and not be interrupted by different arrangements (in theory, not even by free days, such as on weekends). It appears that certain people are more willing and able to conform to such regular "non-day" shift regimens than others.

Health complaints of shift workers are often voiced and suspected, but actual negative effects are difficult to prove. Night-shift workers have, on the average, about half an hour less sleep time than persons who are permanently on day shift. However, Carvalhais, Tepas, and Mahan, 1988, found that persons who permanently work the evening shift (e.g., 4 P.M. to midnight) sleep about half an hour longer than persons on the day shift. Also, persons on the night shift often complain about the reduced quality of sleep they get during the day, with noise often mentioned as particularly disturbing.

Some authors have found statistically significant health complaints as a function of shift work, particularly digestive disorders and gastrointestinal complaints, while other researchers have failed to prove significance (Costa, 1996; Folkhard and Monk, 1985; Kogi, 1996; Monk, Folkard, and Wedderburn, 1996). Altogether, no differences have been found in the mortality of night-shift workers compared to workers in other shifts. However, it appears fairly clear that persons who suffer from health disturbances are more negatively affected by night shifts than by other shift arrangements. It also appears that older workers, possibly due to deteriorating health and difficulties in getting sufficient restful sleep (both phenomena that apparently increase with age) may be more negatively affected by shift work than younger workers.

On the other hand, some older shift workers have relatively fewer health, sleep, and social problems than their younger colleagues, probably due to less need for sleep, a shift to "morningness," and looser and fewer social and family ties. Other aging people suffer from declining health and have more trouble with night work. Haermae (1996) proposes that continuous shift work should be made voluntary for persons over 40 years of age.

Performance

The reduced quantity and quality of sleep experienced by night workers leads to the conclusion that many suffer from a chronic state of partial sleep deprivation. Negative effects of sleep deprivation on behavioral aspects have been well demonstrated, as discussed earlier. For some tasks, the interaction of circadian discrepancies between work demand and body state and of sleep deprivation may result in significant detriments to night work performance, including safety (Monk, 1989). (Accident statistics are usually confounded by many variables in addition to the shift factor, such as work task and worker age and skill).

During the first night shift or shifts, performance is likely to be impaired between midnight and the morning hours, with the lowest performance around 4 A.M. Such impairment, which may be absent or minimal for cognitive tasks, varies in level but is similar to that induced by "legal doses" of alcohol (Monk, 1989). However, as the worker continues to work night shifts, the internal clock realigns itself with the new activity rhythms, and a daily routine of social interactions, sleeping, and going to work are established.

Tolerance for shift work differs from person to person and varies over time. Three out of 10 shift workers have been reported to leave shift work within the first three years as a result of health problems (Bohle and Tilley, 1989). The tolerance of those remaining on shift work depends on personal factors (age, personality, interpersonal troubles, and diseases; the capacity for flexible sleeping habits and for overcoming drowsiness), on social-environmental conditions (family composition, housing conditions, social status) and, of course, on the work itself (workload, shift schedules, income, other compensation). These factors interact, and their importance differs widely from person to person and changes over each work life (Costa et al., 1989; Rosa and Colligan, 1988; and Volger et al., 1988). Evening-shift workers suffer particularly in their social and domestic relations, while night-shift workers are more often affected by the conflicts between the requirement to work while physiologically in a resting stage and by insufficient sleep during the day. However, physiological and health effects are not abundant in shift workers who have been on such assignments for years, possibly because persons who cannot tolerate these conditions abandon shift work soon after trying it.

Social Interaction

Needs for social interaction are individually and culturally different. For example, parents of small children want to be at home for family life and are unlikely to accept unusual work assignments that keep them away, while older persons who do not need to interact with their children so intensely may be more inclined to work "non-normal" hours.

A major problem associated with shift work is the difficulty of maintaining normal social interactions when the work schedule forces one to sleep during times at which social relations usually occur. This makes family relations difficult, as well as interaction with friends and participation in public events such as sports. Common, daily activities, such as shopping or watching television may not be easily carried out.

Care must be taken, however, when transferring one's own living conditions and social expectations to different countries and cultures: certain events or conditions may be present or not, may be regularly scheduled at different times, and be considered of different value to individuals and the society at large. Siesta time is not commonly known in northern regions. In Europe, some shops that are open continuously in the United States are locked around midday, close in the late afternoon, and stay locked on weekends. Family ties are much more important in some cultures than in others and may vary among individuals. Television plays a large role in the daily life of some groups of people. Thus, statements regarding the effects of shift work on social interactions may apply in one case but not pertain to another. It is generally true, however, that shift work and its consequences to the individual worker often interfere with social relations. Whether this has a demonstrable effect on well-being and performance depends on the individual.

How to Select a Suitable Work System

The foregoing discussions should have made it clear that, if at all possible, the working hours should be from morning through afternoon. In many cases, however, this "normal" arrangement is replaced by shift work, occupying either the late afternoon and evening hours, or the night.

There is an apparently inevitable fall in performance during overnight work, related to the circadian rhythm. This is of particular concern with long periods of duty and when the overnight work period follows poor sleep. Reports of reduced performance during the night are numerous, as summarized by Rogers et al. (1989).

The argument for permanent assignment to either an evening or a night shift is well founded on the grounds that such a permanent arrangement allows the internal rhythms to become reentrained according to this rest/work pattern. But that reasoning is not as convincing as it might appear: most shift arrangements are not truly consistent or permanent because the weekend interrupts the cycle. Furthermore, strong zeitgebers (such as light and dark) during the 24-hr day remain intact, even for the person on regular evening or night shifts, thus hindering a complete reentrainment of the internal functions. This leads to the opposite conclusion, also well founded in theory: that it is better to work only occasionally outside the morning/afternoon period and to work only one such evening or night shift. In this case, most people are able to perform their unusual work for this one work period without much detriment while they remain entrained in the usual 24-hr cycle. Of course, some individuals are able and willing to adjust fairly easily to different work patterns. Thus, unusual work patterns may be more acceptable to those who volunteer.

For airplane crews who must cross time zones in long-distance flights and catch some sleep at their destination before returning, several problems exist. The first is that the quality and length of sleep at the stopover location is often much less satisfactory than at home. The resulting tiredness is often masked or counteracted by the use of caffeine, tobacco, and alcohol. The second problem is the extended time of duty, which includes preflight preparations, the flight period itself, and the wrap-up after arrival at the stopover. Negative effects are substantially larger after an eastward flight than a westward one; also, effects on crew members over 50 years old are more pronounced than on their younger colleagues (Graeber, 1988).

Recommendations for the shift arrangement for flight crews are fairly well established. In general, but particularly when flying eastward, flight crews should adhere to well-planned timing. This should duplicate, as far as possible, the sleep/wake activities at home, meaning that the crew should try to go to bed at their regular home time and get up at their regular home time. Thus, they maintain their regular diurnal rhythm. Of course, their next flight duty should be during their regular time of wakefulness.

With respect to the length of a work shift, one general recommendation is that physically demanding work should not be expected over periods longer than 8 hr unless frequent rest pauses are available; but even an 8-hr shift may be too long for very strenuous work. The same applies to work that is mentally very demanding, requiring complex cognitive processes or a high level of attention. For other "everyday" work, durations of 9, 10, or even 12 hours per day can be quite acceptable. Flextime arrangements often are welcomed by employees, possibly in combination with compressed workweeks, particularly if they allow extended free weekends.

As a part of the decision to select one of the many possible shift plans, criteria such as the following must be established to allow justifiable and systematic judgments.

For example:
- Daily work duration should not exceed 8 hr.
- The number of consecutive night shifts should be as small as possible; at best, only one single night shift should be interspersed among the other work shifts.
- Each night shift should be followed by at least 24 hr of free time.
- Each shift plan should contain free weekends, at least two consecutive work-free days.
- The number of free days per year should be at least as large as for the continual day worker.

Using these criteria, Knauth (1996) discussed the design of shift plans that comply with organizational, physiological, and psychological requirements. The selected shift plan will inevitably have drawbacks, but these can be identified and counteracted, possibly by providing special health care services (Koller, 1996).

For evening or night shifts, high illumination levels should be maintained at the workplace, such as 2,000 lux or more. This helps to suppress production of the hormone melatonin, which causes drowsiness. Furthermore, environmental stimuli should be employed to keep the worker alert and awake, such as occasional "stirring" music, provision of hot snacks and of (caffeinated) hot and cold beverages. The work should be kept interesting and demanding since boring and routine tasks are difficult to perform efficiently and safely during the night hours.

The shift worker should use coping strategies for setting the biological clock, obtaining restful sleep, and maintaining satisfying social and domestic interactions. Unless the shift worker is on a very rapidly rotating schedule, the aim is to reset the biological clock appropriately to the shift work regimen. For example, sleep should be taken directly after a night shift, not in the afternoon. Sleep time should be regular and free from interruptions. Shift workers should seek to gain their family's and friends' understanding of their rest needs. Certain times of the day should be set aside specifically and regularly to be spent with family and friends.

SUMMARY

Human body functions and human social behavior follow internal rhythms. Aside from the female menstrual cycle, the best-known rhythms are a set of daily fluctuations called circadian or diurnal rhythms. Examples are body temperature, heart rate, blood pressure, and hormonal excretions. Under regular living conditions, these temporal programs are well established and persistent.

The well-synchronized rhythms and the associated behavior of sleep (usually during the night) and of activities (usually during the day) can be desynchronized and put out of order if the time markers (zeitgeber) during the 24-hr day are changed and if activities are required of people at unusual times. Resulting sleep loss and tiredness influence human performance in various negative ways. Mental performance, attention, and alertness usually are reduced, but most physical activities are not affected. Furthermore, concerns exist that disturbing the internal rhythm, especially by night-shift work, might have negative health effects. Certainly, being excluded by evening-shift work, for example, from participating in family and social activities is difficult for many persons.

Shift work is often desired for organizational/economic reasons. Individual acceptance of working in shifts depends on a complicated balance of professional and personal concerns, including physiological, psychological, and social aspects. Of 10 persons assigned to shift work, 7 or 8 are likely to stay on this schedule, while the others drop out.

Preexisting health problems may be exacerbated by shift work. Gastrointestinal/digestive problems are fairly frequent among shift workers. Persons on permanent night shifts often complain about insufficient sleep and general fatigue.

As for work performance of shift workers, or accidents, little reliable information is available because, on evening and night shifts, different work is often performed than that done during the day. But work (other than cognitive) on a person's first night shift is likely to be impaired in the early morning hours by "fatigue" that is similar to the effects of small doses of alcohol.

Based on a review of physiological, psychological, social, and performance aspects, recommendations for acceptable regimes of working hours and shift work include:

Job activities should follow entrained body rhythms.
It is preferable to work during the daylight hours.
Evening shifts are preferred to night shifts.

If shifts are necessary, two opposing rules apply:

(1) Either work only one evening or night shift per cycle, then return to day work, and keep weekends free; or (2) stay permanently on the same shift, whatever that is.

A shift duration of 8 hr of daily work is usually adequate, but shorter times for highly (mentally or physically) demanding jobs may be advantageous; longer times (such as 9, 10, or even 12 hr) may be acceptable for some types of routine work.

Compressed workweeks often are acceptable for routine jobs.

REFERENCES

Basmajian, J.V., and DeLuca, C.J. (1985). *Muscles Alive,* 5th ed. Baltimore, MD: Williams & Wilkins.

Bohle, P., and Tilley, A.J. (1989). The Impact of Night Work on Psychological Well-Being. *Ergonomics* 34(9):1089–1099.

Carvalhais, A.B., Tepas, D.I., and Mahan, R.P. (1988). Sleep Duration in Shift Workers. *Sleep Research* 17:109.

Colligan, M.J., and Tepas, D.I. (1986). The Stress of Hours of Work. *American Industrial Hygiene Association Journal* 47(11):686–695.

Colquhoun, W.P. (1985). Hours of Work at Sea: Watch-keeping Schedules, Circadian Rhythms and Efficiency. *Ergonomics* 28(4):637–653.

Costa, G. (1996). The Impact of Shift and Night Work on Health. *Applied Ergonomics* 27(1):9–16.

Costa, G., Lievore, F., Casaletti, G., Gaffuri, E., and Folkard, S. (1989). Circadian Characteristics Influencing Inter-Individual Differences in Tolerance and Adjustment to Shiftwork. *Ergonomics* 32(4):373–385.

Dunham, R.B., Pierce, J.L., and Castaneda, M.B. (1987). Alternative Work Schedules: Two Field Quasi Experiments *Personnel Psychology,* 40:215–242.

Eastman Kodak Company (ed.) (1986). *Ergonomic Design for People at Work, Vol. 2.* New York, NY: Van Nostrand Reinhold.

Folkard, S., and Monk, T.H. (eds.) (1985). *Hours of Work.* Chichester, UK: Wiley.

Froeberg, J.E. (1985). Sleep Deprivation and Prolonged Working Hours. Chapter 6 in S. Folkard and T.H. Monk (eds.). *Hours of Work.* Chichester, UK: Wiley, pp. 67–76.

Gillberg, M. (1985). Effects of Naps on Performance. Chapter 7 in S. Folkard and T.H. Monk (eds.). *Hours of Work.* Chichester, UK: Wiley, pp. 77–86.

Graeber, R.C. (1988). Aircrew Fatigue and Circadian Rhythmicity. Chapter 10 in E.L. Wiener and D.C. Nagel (eds.). *Human Factors in Aviation.* San Diego, CA: Academic Press, pp. 305–344.

Haermae, M. (1996). Ageing, Physical Fitness and Shiftwork Tolerance. *Applied Ergonomics* 27(1):25–29.

Horne, J. (1988). *Why We Sleep—The Functions of Sleep in Humans and Other Mammals.* Oxford, UK: Oxford University Press.

Hunter, K.I., and Shane, R.H. (1979). Time of Death and Biorhythmic Cycles. *Perceptual and Motor Skills* 48(1):220.

Hurrell, J.J., and Colligan, M.J. (1985). Alternative Work Schedules: Flextime and the Compressed Workweek. Chapter 8 in C.L. Cooper and M.J. Smith (eds.). *Job Stress and Blue Collar Work.* New York, NY: Wiley, pp. 131–144.

Johnson, L.C., Tepas, D.I., Colquhoun, W.P., and Colligan, M.J. (eds) (1981). *Biological Rhythms, Sleep and Shift Work.* New York, NY: Spectrum.

Kerkhof, G. (1985). Individual Differences and Circadian Rhythms. Chapter 3 in S. Folkard and T.H. Monk (eds.). *Hours of Work.* Chichester, UK: Wiley, pp. 29–35.

Knauth, P. (1996). Designing Better Shift Systems. *Applied Ergonomics* 27(1):39–44.

Kogi, K. (1985). Introduction to the Problems of Shiftwork. Chapter 14 in S. Folkard and T.H. Monk (eds.). *Hours of Work.* Chichester, UK: Wiley, pp. 115–184.

Kogi, K. (1996). Improving Shift Workers' Health and Tolerance to Shiftwork: Recent Advances. *Applied Ergonomics* 27(1):5–8.

Koller, M. (1996). Occupational Health Services for Shift and Night Workers. *Applied Ergonomics* 27(1):31–37.

Lateck, J.C., and Foster, L.W. (1985). Implementation of Compressed Work Schedules: Participation and Job Redesign as Critical Factors for Employee Acceptance. *Personnel Psychology* 38:75–92.

Lavie, P. (1985). Ultradian Cycles in Wakefulness. Chapter 9 in S. Folkard and T.H. Monk (eds.). *Hours of Work.* Chichester, UK: Wiley, pp. 97–106.

Minors, D.S., and Waterhouse, J.M. (1981). *Circadian Rhythms and the Human.* Bristol, UK: Wright.

Monk, T.H. (1989). Shiftworker Safety: Issues and Solutions. In A. Mital (ed.). *Advances in Industrial Ergonomics and Safety I.* Philadelphia, PA: Taylor & Francis, pp. 887–893.

Monk, T.H., Folkard, S., and Wedderburn, A.I. (1996). Maintaining Safety and High Performance on Shiftwork. *Applied Ergonomics* 27(1):17–23.

Monk, T.H., and Tepas, D.I. (1985). Shiftwork. Chapter 5 in C.L. Cooper and M.J. Smith (eds.) *Job Stress and Blue Collar Work.* New York, NY: Wiley, pp. 65–84.

Patkai, P. (1985). The Menstrual Cycle. Chapter 8 in S. Folkard and T.H. Monk (eds.). *Hours of Work.* Chichester, UK: Wiley, pp. 87–96.

Persinger, M.A., Cooke, W.J., and Janes, J.T. (1978). No Evidence for Relationship Between Biorhythms and Industrial Accidents. *Perceptual and Motor Skills,* 46(2):423–426.

Rechtschaffen A., and Kales, A. (1968). *A Manual of Standardized Terminology, Techniques, and Scoring System of Sleep Stages in Human Subjects.* Los Angeles, CA: UCLA Brain Information Services.

Rogers, A.S., Spencer, M.B., Stone, B.M., and Nicholson, A.N. (1989). The Influence of a 1h Nap on Performance Overnight. *Ergonomics* 32(10): 1193–1205.

Rosa, R.R., and Colligan, M.J. (1988). Long Workdays Versus Rest Days: Assessing Fatigue and Alertness with a Portable Performance Battery. *Human Factors* 30(3):305–317.

Tepas, D.I. (1985). Flextime, Compressed Workweeks and Other Alternative Work Schedules. Chapter 13 in S. Folkard and T.H. Monk (eds). *Hours of Work.* Chichester, UK: pp. 147–164.

Tepas, D.I., and Monk, T.H. (1986). Work Schedules. Chapter 7.3 in G. Salvendy (ed). *Handbook of Human Factors*. New York, NY: Wiley Interscience, pp. 810–843.

Volger, A., Ernst, G., Nachreiner, F., Haenecke, K. (1988). Common Free Time of Family Members in Different Shift Systems. *Applied Ergonomics* 19(3): 213–228.

Wever, R.A. (1985). Men in Temporal Isolation: Basic Principles of the Circadian System. Chapter 2 in S. Folkard and T.H. Monk (eds). *Hours of Work*. Chichester, UK: Wiley, pp. 15–28.

FURTHER READING

Applied Ergonomics 27:1, 1996. (Several articles.)

Folkard, S., and Monk, T.H. (eds) (1985). *Hours of Work*. Chichester, UK: Wiley.

Johnson, L.C., Tepas, D.I., Colquhoun, W.P., and Colligan, M.J. (eds) (1981). *Biological Rhythms, Sleep and Shift Work*. New York, NY: Spectrum.

Horne, J. (1988). *Why We Sleep—The Functions of Sleep in Humans and Other Mammals*. Oxford, UK: Oxford University Press.

Chapter 11

The Reengineered Human Body:
A Look into the Future

OVERVIEW

The human body is a highly complex and internally interdependent organism. However, the functioning of elements of the body can be analyzed from an engineering point of view and may often, with varying degrees of success, be mimicked. This analysis is important to better understand and interpret function, to aid healing and rehabilitation and, in some cases, to replace diseased or nonfunctioning portions of the body.

THE MODEL

The body, like other complex engineering structures, is constructed of many different parts with distinct functions. Using techniques from mechanics and biology, we can attempt to understand and replicate some of the parts and functions of the body.

INTRODUCTION

With all the wonderful techniques of modern science, we are just now approaching an understanding of the complexity of the human body. It is a highly interdependent internal system: a malfunction in one biologic component can affect the functioning of other components. This may be as simple as the pain from a partially torn ligament in the knee causing a person to change her gait to compensate. Or it may be as complex as a sprained back muscle influencing posture, which may accelerate disk degeneration, leading to back pain that causes the person to change most aspects of his daily living.

Engineering models of various aspects of human physiology tend to ignore biologic and chemical workings of the body, especially at the cellular level.

However, even the relatively coarse models used so far have allowed technical advances that can reduce pain and increase function.

The human body has been reengineered at many levels, even though the one-time television shows *The Six Million Dollar Man* and *The Bionic Woman* or the "morphed" beings remain fantastic fiction. In reality, the collaboration between medicine, biology, and engineering has led to advances in the development of artificial joints, organs (including skin), teeth, limbs (prosthetics), blood vessels, and even augmenting senses. For the purposes of this chapter, these subjects are loosely grouped by their function into replacements and augmentation for parts of the human body.

This chapter is, of course, incomplete. The reengineering of the human body is ever ongoing. For further details on these evolving topics, stay tuned to the literature, some of which is referenced at the end of this chapter.

BASIC CONCEPTS FOR HUMAN BODY DESIGN

Those reengineered components for the human body that come into intimate contact with body fluids need to satisfy rigorous criteria. The requirements differ in their severity depending on the exact application, where they are used, how easy it is to replace the device, and how severe the consequences of failure are. For example, if an artificial heart valve causes infection, the resulting blood clot could kill the patient; thus, a candidate valve must be designed with great care. The consequences of an external hearing aid failing might be less life-threatening but, nevertheless, demands must be met regarding reliability, usefulness, appearance, and so forth.

Basic requirements for design of replacement and augmentative components include the following:

- The materials used must be biocompatible.
- The design must be implantible (if applicable).
- The component must have a useful service life.
- There must be a way to access the component in case it fails.
- The device should not be cumbersome and should be easy to use.

Biocompatibility encompasses a variety of materials issues for the candidate material and for the biologic tissues with which it will interact. The material must be compatible with the host tissue in terms of physiology and chemistry (and even electricity for muscle-related implants), both in its bulk form and as particles that may be shed from the material while in use. The material must be able to withstand the loads expected to be placed on it for the length of time that it might be expected to be used and must have a stable and/or predictable response to its service use. Further, the material must be able to withstand the environment in which it is placed: body fluids may be warm and corrosive to many types of materials. Not least, the material must be able to be sterilized if it is to be implanted.

Interfaces are a further material consideration, concerning both the interfaces between artificial components, which is a conventional engineering aspect, and the interfaces between the component and the host tissue.

These criteria overlap with the following criteria which are common in rehabilitation engineering (Kroemer, Kroemer, and Kroemer-Elbert, 1994). The device must have

- Affordability—the extent to which the purchase, maintenance, and repair causes financial hardship to the consumer.
- Dependability and durability—the extent to which the device operates with repeatable and predictable levels of accuracy for extended periods of time.
- Physical security—the probability that the device will not cause physical harm to the user or to other people.
- Portability—the extent to which the device can be readily transported to, and operated in, different locations.
- Learnability and usability—the extent to which the consumer can easily learn to use a newly received device and can use it easily, safely, and independently for its intended purpose.
- Physical comfort and personal acceptability—the degree to which the device provides comfort or, at least, avoids pain or discomfort to the user, so that the consumer is willing to use it in public or private.
- Flexibility and compatibility—the extent to which the device can be augmented by options and to which it will interface with other devices used currently or in the future.
- Effectiveness—the extent to which the device improves the user's capabilities, independence, and objective and subjective situation.

REPLACEMENTS FOR BODY PARTS

Artificial Joints

Total replacement for major joints in the body has become quite popular: hundreds of thousands of hips and knees are replaced each year in the United States alone. Joint replacement has become one of the most successful surgeries to ease pain and restore function to a disabled patient. Success has led to the development of replacements for many joints, including the shoulder, elbow, wrist, fingers, hip, knee, and ankles. The most popular and generally the longest-lasting replacement joints are the hip and knee. (See Chap. 2 for more details.)

Total replacement of joints requires invasive surgery. Often this is undertaken for osteoarthrosis, which is a wearing away of the joints due to trauma, cumulative use or abuse, and certain diseases. The cartilage surfaces of the bones at the joints are damaged and generally are removed at the time of total joint replacement surgery. Sufficient bone is also removed to allow anchoring of the components that make up the new joint surfaces. For the hip joint, the acetabulum in the pelvic bone is resurfaced with a new, plastic cup. The head of the femur (thigh bone) is removed and replaced by a metallic sphere mounted on a stem that fits into the femoral canal. For the knee joint, the condyles of the femur are resurfaced with a metallic component while a plastic tray is placed on the flattened top of the tibia (shinbone). Often, the patella (kneecap) is also resurfaced with a plastic button.

Long-term success rates for hip and knee replacements are typically greater than 90% at 10 to 15 years. However, revision of the primary replacement may become unavoidable in some cases if the patient experiences excessive pain due to infection of the joint, loosening of one or more of the components in its bony bed, or unacceptable wear of the articulating surfaces. To replace the artificial joint components requires a second surgery; success rates following revision surgery are not as high as those following the primary surgery. However, given the overall success rate of the procedure and its dramatic relief of pain, it is not surprising that replacement of hips and knees has become common.

Artificial Teeth (Dental Implants)

For those who have lost some or all of their teeth, endosteal dental implants are often indicated. These are artificial teeth that are implanted into the mandibular (jaw) or maxillary bone. Either a one-stage or two-stage procedure may be used. For a one-stage device, the base of the artificial tooth is implanted into the bone. For a two-stage device, an anchor is first placed in the bone, then left for some time (maybe months) to heal. Then the tooth prosthesis is placed onto the now firmly ingrown anchor. The anchor may be made from a titanium alloy or a ceramic (such as alumina oxide). To aid in fixing the anchor to the bone, a screw thread or even a fin design may be used. The anchors may also be used to secure more elaborate dental implants, such as bridges and dentures.

Artificial Prostheses (Limbs)

Partial and whole limb prosthetics are designed for amputees and others with malformed or missing limbs, typically to mimic the function and/or the appearance of arms and legs. The conflicting design objectives for limb prosthetics include modularity (or other approach to permit use by those with varying deficits), function, cost, weight, and size. Prosthetics may be unpowered, manually operated, or driven by a motor fueled by electricity or pneumatics. One recent arm prosthetic allows motion at the hand, wrist, elbow, and shoulder, while having a mass of only 2 kg, including batteries; this arm is capable of lifting more than one kg at full arm extension (Dow, 1996). Another contemporary design for an above-the-knee leg prosthetic uses microcomputers to control gait through sensors that detect outputs from the remaining muscles at the stump, as well as the knee angle of the prosthesis and pressure at the foot of the prothesis (Van Petegem, Vander Sloten, Van Audekercke, and Van der Perre, 1996).

AUGMENTATION TO EXISTING BODY COMPONENTS

Artificial Ligaments

The ligaments of the knee are prone to injury due to the large loads and inherent instability of the bony geometry of the knee. The anterior cruciate ligament (ACL) is the one most commonly torn, either partially or fully. If this ligament is no longer functioning, the knee may sublux (dislocate) anteriorly or posteriorly and eventually may experience damage to the other components in the joint. Reconstruction of the ligaments of the knee date back at least to the 1930s. Commonly, the ACL is replaced using a graft from the same patient or a donor, from either the patellar tendon or another structure. However, failures result from biologic problems, insufficient strength of the graft, poor placement of the graft, and graft fixation. Complications may also arise at the graft site if donor material from the patient is used. There are ongoing efforts to develop an artificial ligament instead of a damaged natural ligament or to augment the function of the natural ligament: this avoids graft site complications and risks inherent in grafts from other donors. However, problems with synthetic ligaments include issues of biocompatibility, ligament strength, long-term functioning of the device, and weakening of the ligament fixation to the host bone (see, for example, Müller, 1996).

Artificial Arteries

Materials to be used in the vascular system have special requirements, among them that they must not excite an excessive response from the blood itself. Therefore, one must consider carefully the surface that is exposed to blood.

Grafts are used to bypass portions of arteries that are partially or totally blocked or that, in aneurysms, expand like balloons. Currently, there are no manufactured grafts acceptable for coronary arteries, so that human vein grafts are commonly used. The development of small-diameter (4-mm or less) vascular grafts has included attempts using biologic sources as well as plastics, such as polyester, polytetraflouroethylene (PTFE), polyurethane, and a combination of a graft vein augmented with polyethylene terephthalate (Dacron®) (Szycher, 1991; Sharma, 1991). However, in attempts to permanently replace arteries, problems persist with blood clot formation.

In some cases, arteries are cleaned or opened using a balloon catheter in an angioplastic procedure. The catheter is threaded to the blockage from some remote site (such as to a blocked coronary artery from the femoral vein in the groin). Using fluoroscopic x-rays to guide the catheter, the balloon tip is inflated once the blocked site is reached. To help the reinflated vessel remain open, a stent may be inserted. In the 1980s, metallic stents were tried: stainless steel or tantalum were used as these metals have been previously shown to be biocompatible. Because of less than satisfactory results with these stents, attributed in part to the metal's net electrical charge or potential (affecting the blood's response) and free surface energy, research is under way for new stent materials. Current work is considering novel geometries (tubes and meshes), materials, and therapeutic coatings. There are even attempts to use biodegradable polymers that would be absorbed into the body over time (Peng et al., 1996).

Artificial Organs

Skin

In the case of severe and widespread wounds and skin burns, one needs to be concerned about loss of body fluids and scar tissue. To avoid these problems, special dressings and artificial skins are used. There have been attempts to use biologic substances, albumin and collagen, onto a polyvinyl alcohol (PVA) substrate, but clinical success is still in the future (Aleyamma and Sharma, 1991).

Lung

The lung is essentially a mass exchanger, where the blood is enriched with oxygen and carbon dioxide is removed (see Chap. 6). Artificial lungs (also called oxygenators) are most commonly used in cardiac surgery to temporarily bypass the natural heart and lungs; temporarily, in this case, means several hours. These oxygenerators often expose the blood to the ventilating gas (direct contact). Artificial lungs for longer-term use generally have a permeable membrane between the blood and the gas used to oxygenate the blood (indirect contact). Of course, the natural lung also performs tasks other than gas exchange, such as blood filtration, enzyme production, and influence over hemostasis and infection. Artificial lungs cannot perform all these functions and usually concentrate only on maintaining physiologic levels of oxygen and carbon dioxide in the blood (Gaylor, 1980).

Kidney

Like the lung, the kidney is a mass exchanger, where wastes dissolved in the blood are removed across a semipermeable membrane. In cases of kidney failure (whether acute or chronic), cleansing of the blood is generally performed outside the body: extracorporeal hemodialysis. For those with chronically nonfunctioning kidneys, hemodialysis can be undertaken as often as necessary (generally, two to three times per week for several hours each time) for many years. This maintenance dialysis requires continual invasion of the bloodstream; often, external or internal shunts are placed in the arms or legs to allow easy access to the blood (Gaylor, 1980). Polymer membranes are the fundamental component of artificial kidneys. They transport the blood to and from the exchange site and often make up the waste exchange mechanism itself. Cellulose, cellulose acetate, and cellulose nitrate filters in various shapes have all been used for hemodialysis (Courtney and Gilchrist, 1980). The natural kidney is responsible for removal of wastes (especially urea and other products of protein metabolism), for the regulation of acid-base, electrolyte balance, and fluid volume, for partial hemodynamic control of pressure and volume, and some hormonal influences. Artificial kidneys address only the removal of wastes and, partially, the regulatory functions (Gaylor, 1980).

Heart

Heart dysfunctions may be due to mechanical problems with the valves, irregularity of pumping rhythm, and poor performance of the pumping action of the heart. These are treated with different procedures and devices, such as replacement of the total heart with an artificial or a donor heart (first done with some success in the 1960s), replacement of valves only, or artificial pacing to provide assistance to the heart's internal pacemaker (see Chap. 7).

The valves in the heart are designed to allow blood flow in only one direction, from the atrium to the ventricle to the artery, without allowing flow in the reverse direction (called *retrograde flow*). Diseased valves, which are generally on the left side, are often replaced with prosthetic valves (Dellsperger and Chandran, 1991). The prosthetic valves may be made from metal, from plastic, or from biologic sources (especially from pigs); some are bioprosthetics, made from a metal framework onto which a biologic tissue (such as the pericardium from a cow) is sewn.

When there are problems with the heart's electrical system that restrict its ability to pump sufficient blood, a heart may be assisted using pacing systems. An artificial pacemaker can be used to stimulate the heart to pump appropriately. The right and left side of the heart contract together, so that only one side (generally, the right) need be stimulated artificially. Pacemakers generally weigh less than 100 g and last from 5 to 10 years or more (Szycher, 1991).

If a heart is not putting out enough blood at a sufficient pressure, the cardiac output is inadequate: this defines congestive heart failure (CHF). Drugs may be used with some success but, when these are not sufficient, a heart transplant may be the best solution. There have been some high-profile attempts in the United States to use nonhuman hearts, such as from a baboon, or even an artificial heart that pumps the blood using external machinery. Artificial hearts can be either ventricular assist devices, where the natural heart remains in place, but the device is attached between the left ventricle and the aorta, in parallel with the natural heart; or they can be a total heart replacement when the natural heart is removed (Szycher, 1991). Emergency-use assist devices, such as an intra-aortic balloon pump or a temporary left ventricular assist device (LVAD) or biventricular assist device, are meant to ease the strain on a failing heart to give it time to recover but are not for long-term use. Currently, no artificial hearts are practical for long-term use. This is due both to material issues of stability and biocompatibility and to infection. Instead, artificial hearts may be used as bridge devices until a donor human heart becomes available: waiting periods are on the order of a month (Kambic and Nose, 1991). In the United States alone, over a thousand heart transplants are performed each year. Because of a shortfall in the number of donor hearts available, many more patients die while waiting for an appropriate heart to become available.

Artificial Senses (Augmentation): Seeing and Hearing Aids

Aids to the hearing and seeing senses have a long tradition: a funnel held to the ear "to catch the sound" and glass lenses to "enlarge the object" are shown in medieval illustrations. These attempts to overcome reduced capabilities, often part of the aging process, have become more effective with new technologies.

To improve hearing or to provide protection against harmful sounds, the specific parts and functions of the ear can be treated by engineering means. Occlusion of the outer ear to keep excessive sound energy away is done by helmets and muffs ("caps") that encompass the ear lobe or by inserts ("plugs") into the ear canal. New developments go beyond the simple damming of sound but also amplify those sounds of the acoustic environment that one wants to hear, for example, as part of conversation, or must hear such as a warning signal embedded in noise. Other current sound-deadening devices use active elimination by instantly generating a similar sound that is 180 deg off-phase. Until recently, not much could be done to help persons who had suffered damage to the sound receptors in the inner ear except filtering and amplifying incoming sound, but successful attempts have been made to replace the cochlea with an electronic implant that picks up the incoming signals and transforms them so that they can be transmitted along the acoustic nerve to the brain.

To help deteriorated vision, the "focusing system" (consisting primarily of cornea and lens) can be improved by engineering means. Probably the most common eye disorder is near-sightedness (myopia); one cannot focus on far objects. This is often the result of a cornea whose curvature bulges too much so that light rays focus in front of the retina rather than directly on it. Another often occurring condition of the cornea is uneven curving (astigmatism), which prevents clear focus on any visual target. Both deficiencies usually can be overcome by the traditional technique of putting artificial lenses in front of the eye in the form of either spectacles or contact lenses. These lenses are formed to compensate for the imperfect shape of the cornea. A newer technique, called radial keratotomy, reshapes the cornea itself: incisions are made that radiate from the center of the cornea. The incisions, originally made by scalpel and now usually by laser, flatten the cornea. The operation is often highly successful, takes only a few minutes, and is painless.

Another successful operation is the total replacement of the natural lens by an artificial one: a fast and painless operation that is helping many older persons who have incurred cataracts that make the lens opaque, a condition that, in the past, has led to blindness.

SUMMARY

In the engineering view, the human body is an exceedingly complex structure with numerous parts that serve specific purposes and that are interdependent in many of their functions. Taking one part at a time, and considering only the most important interfaces with the rest of the body, one is, in some cases, able to replicate all, most, or just a few of its functions; helping people to do things better than would be possible without the engineering intervention. Great advances have been made: in rehabilitation engineering, consider the humble start, with a simple wooden peg strapped to the stump of a leg, to articulated and powered artificial legs, to today's powered prostheses, controlled by computer and nerve signals picked up in the intact efferent part of the peripheral nervous system. In biomedical engineering, mass-manufactured hip and knee joints, artificial ligaments, manufactured skin, plastic eye lenses, and electronic cochlea implants are at hand. Much can be expected from genetically engineered organs, grown inside or outside the human body. Exciting engineering topics indeed!

The first 10 chapters of this book provided background information for the human factors engineer and manager to understand the functions of the human body so that

Work processes, jobs, and specific tasks
Workstations, tools, equipment, and machinery

Shops and offices
Work environment and protective clothing
Shift and working time

can be laid out according to the ergonomic axiom, "Make work easy and efficient."

REFERENCES

Aleyamma, A.J., and Sharma, C.P. Polyvinyl Alcohol as a Biomaterial. Chapter 6 in C.P. Sharma and M. Szycher (eds.) (1991). *Blood Compatible Materials and Devices: Perspectives Towards the 21st Century.* Lancaster, PA: Technomic Publishing.

Courtney, J.M., and Gilchrist, T. "Polymers in Medicine." In R.M. Kenedi (ed.) (1980). *A Textbook of Biomedical Engineering.* London: Blackie & Son.

Dellsperger, K.C., and Chandran, K.B. "Prosthetic Heart Valves." Chapter 9 in C.P. Sharma and M. Szycher (eds.) (1991). *Blood Compatible Materials and Devices: Perspectives Towards the 21st Century.* Lancaster, PA: Technomic Publishing.

Dow, D.J. 1996. Prosthetic Upper Limb Research in Edinburgh. *10th Conference of the European Society of Biomechanics, Leuven, August 28–31, 1996;* ed. J. Vander Sloten, G. Lowet, R. Van Audekercke and G. Van der Perre, p. 82.

Gaylor, J.D.S. Artificial Organs. In R.M. Kenedi (ed.) (1980). *A Textbook of Biomedical Engineering.* London: Blackie & Son.

Kambic, H.E., and Nose, Y. Biomaterials for Blood Pumps. Chapter 8 in C.P. Sharma and M. Szycher (eds.) (1991). *Blood Compatible Materials and Devices: Perspectives Towards the 21st Century.* Lancaster, PA: Technomic Publishing.

Kroemer, K.H.E., Kroemer, H.B., and Kroemer-Elbert, K.E. (1994). *Ergonomics: How to Design for Ease and Efficiency.* Englewood Cliffs, NJ: Prentice-Hall.

Müller, W. (1996). Knee Ligament Injuries: Pathoanatomy, Biomechanics, Instabilities and Possibilities of Treatment in Acute and Chronic Injuries. *International Orthopedics (SICOT)* 20:266–270.

Peng, T., Gibula, P., Yao, K., and Goosen, M.F.A. (1996). Role of Polymers in Improving the Results of Stenting in Coronary Arteries. *Biomaterials* 17(7):685–694.

Sharma, C.P. Surface Modification: Blood Compatability of Small Diameter Vascular Graft. Chapter 3 in C.P. Sharma and M. Szycher (eds.) 1991. *Blood Compatible Materials and Devices: Perspectives Towards the 21st Century.* Lancaster, PA: Technomic Publishing.

Szycher, M. Biostability of Polyurethane Elastomers: A Critical Review. Chapter 4 in C.P. Sharma and M. Szycher (eds.) (1991). *Blood Compatible Materials and Devices: Perspectives Towards the 21st Century.* Lancaster, PA: Technomic Publishing.

Van Petegem, W., Vander Sloten, J., Van Audekercke, R., and Van der Perre, G. 1996. Optimized Power Management for a Microcomputer Controlled Above-

Knee Prosthesis. *10th Conference of the European Society of Biomechanics, Leuven, August 28–31, 1996,* ed. J. Vander Sloten, G. Lowet, R. Van Audekercke and G. Van der Perre, p. 84.

FURTHER READING

Fisk, A.D., and Rogers, W.A. (eds) (1997). *Handbook of Human Factors and the Older Adult.* San Diego, CA: Academic Press.

Kenedi, R.M. (ed.) (1980). *A Textbook of Biomedical Engineering.* London: Blackie & Son.

Kroemer, K.H.E., Kroemer, H.B., and Kroemer-Elbert, K.E. (1994). *Ergonomics: How to Design for Ease and Efficiency.* Englewood Cliffs, NJ: Prentice-Hall.

Sharma, C.P., and Szycher, M. (eds.) (1991). *Blood Compatible Materials and Devices: Perspectives Towards the 21st Century.* Lancaster, PA: Technomic Publishing.

The following list of journals may be helpful to follow progress in the field:

Applied Ergonomics
Biomaterials
Biomechanical Engineering
Clinical Orthopedics and Related Research
Ergonomics
Human Engineering in Design
Human Factors
Journal of Biomechanical Engineering
Journal of Biomechanics
Journal of Biomedical Materials Research
Journal of Bone and Joint Surgery
The Knee
Prosthetics and Orthotics International
Rehabilitation and Progress Reports
Rehabilitation and Research Development
Spine

Index

A-band, 85, 86
abduct, definition of, 57
abduction, 166
absorption, pathways of, 201
acceleration, 104, 105
   definition of, 126
acclimatization, 251–52
acromion, definition of, 57
actin, definition of, 126
actin filaments, 84, 85
action, definition of, 126
action potential, 135
active contractile tension, 88, 96
adduct, definition of, 57
adduction, 166
adenosine disphosphate. *See* ADP
adenosine triphosphate. *See* ATP
ADP (adenosine diphosphate)
   breakdown of, 204
   and muscle work, 206
aerobic metabolism, 206–8
   of glucose, 202
afferent feedback loop, 98
age, 23, 63–64, 69, 78, 177, 256, 271, 286–87, 289
aging. *See* age
agonist muscle, 82–83
air humidity, 242–43, 257
air movement, 243, 257
"all-or-none" principle, 92–93
alveoli, 174, 175
amino acid, 200–201, 203–4
anabolism, 201
anaerobic metabolism, 206–8

of glucose, 202–3
anatomical landmarks, 3–5
anatomical position, 3
anatomical terms, glossary of, 57–60
ancestry, 22
anemometer, 243
anisometric, 82, 102
anisotropic band. *See* A-band
ankle
   maximal displacement of, 67
   mobility of, 50
antagonist muscle, 82–83
anterior, definition of, 57
anthropometer, 4, 6, 8
anthropometric data, 11–12
   age, changes with, 23
   biomechanical functions of, 143
   for body dimensions, 32–38
   body proportions, 14–15
   body segment mass, 161–63
   body volumes, 152, 156
   of civilian populations, 15, 17
   correlations of, 13–14, 15, 16
   data variations in, 25
   estimating, 38–47
   inertial properties, 156–57
   interindividual variations in, 19–22
   intraindividual variations in, 23
   links and joints, 151–52, 153–55
   locating center of mass, 159
   normality, 12–13
   secular variations in, 23–25
   variability in, 13, 44